Intriguing Astronomical Scientific Information

Ronald J. Weinmann

Copyright © 2018 Ronald J Weinmann

All rights reserved.

ISBN-13: 978-1985676930

… Ronald J. Weinmann

Acknowledgments

The first and foremost person I would like to thank for typing all of the handwritten pages of my book is my brother Daniel. Without his dedication in helping me, this book would not have been possible. It was my brother who helped me get my book published as well. Thank you Daniel for everything that you did to make this possible for me.

Let me tell you a bit about myself. I was injured in the year 2000, when I was hit by a drunk driver. I sustained a TBI from the crash. The next five years was spent recovering enough that I could go back to college again. It was a rough five years of physical, occupational and speech therapies. At this time it was my mom who inspired me to have the confidence to start back up at college. Thank you mom. I graduated in 2011 with an Associate of Science degree. I would not have made the accomplishment of writing this book without my mom's help in driving me back to school to further my education.

Another person who has helped me tremendously throughout my life is my dad. Thank you dad for making me tough it out and not quit wrestling. For it was that challenging sport that taught me to have the strong willpower to overcome adversities in my life. Also, thank you

Intriguing Astronomical Scientific Information

dad for helping me use my first telescope in my youth. That sparked an interest that has lasted my whole life.

Table of Contents

- Introduction……………………..………………..………..1
- Hubble Telescope…………………………………..…..4
- Intriguing Astronomical Scientific Information……..6
- Stars……….. ………………………………………..…...69
- The Milky Way Galaxy……………………………...76
- Galaxies…………………………………………………80
- Dark Matter……………………………………….…...87
- The Local Group……………………………………..…...92
- Light-Years (ly)……………………………………………..95
- Super Clusters……………………………… ………....97
- Constellations……………………………………….....100
- Time Travel………………………………………...106
- Black Holes……… …………………………………...108
- The Sun…………………………………………………110
- Trappist-1……………………………………………..112
- Conclusion …………………………………………....115
- Works Cited Page.. ……………………………………119

Introduction

Within this book I will make an attempt of presenting the universe to you. This book will consist of information that I was taught about while receiving my associates degree of science. I will also include evidence that I have acquired upon by learning about outer space. Initially I begin by defining certain words and important terminological phrases, that are decisive when thinking about outer space and the universe that surrounds it.

I will also discuss several space probes and some of the knowledge our scientists at NASA plan on attaining from them. I will take you with me on a hypothetical spaceship and we will travel from our solar system out into the universe. We will look at how our sun is just one star amongst our home galaxy. We will also look at how many other galaxies that we currently know to be in existence. I will describe black holes in detail and the purpose they serve in the universe. I will discuss how we have categorized the distribution of matter in the universe. I'll also clearly define light speed and how space is so immense that it is the only justifiable unit of measurement to represent the vast distances that exist in outer space.

Intriguing Astronomical Scientific Information

I will then expand out into our own galaxy and describe some of its intriguing features. These will be features like the size, shape, how it formed and certain characteristics it has. I will mention the constant motion it possesses.

Next I will touch on the other galaxies in the universe. I will mention their different sizes, shapes and how these galaxies are distributed throughout the universe. I will also include the names of different types of galaxies. I will inform you about the different ways the stars in the galaxies are distributed.

Then I will mention how the dispersion of galaxies in the universe have been grouped together, these groups form clusters. The galaxy clusters are then combined to form super clusters.

I will discuss the different forms of light and heat the universe obtains. I will try to describe dark matter to you. Along with this dark matter there is also dark energy. Each of these particles are difficult to describe because they are invisible. I will describe how we even thought anything was there if we could not see it. I will speak about the expansion of the universe and what it will be believed to cause.

From here on our journey we will take a look at our local group of galaxies the Milky Way belongs to. I briefly will discuss the sizes of some of these galaxies a long with their distances in light years. I will tell you what galaxies can be seen in our night sky. I will clearly define what a light year is and the answer may surprise you.

I will tell you about how we have categorized the dispersion of matter in the visible universe from galaxies, to groups of galaxies. Next the groups are combined to form clusters. Finally the clusters are united to form super clusters. I will also tell you how many super clusters we have found to be in existence. I will then show you some of my favorite constellations in the night sky with the names of their stars and the distances to them in ly's.

I briefly touch on time travel and mention several ways that some believe it to be possible. I then will tell you about black holes, what they are, what they do and how they affect things that are around them. I will tell you where every piece of matter we know to be in existence lies. I then will mention several scenarios as to what is in store for the existence of our universe.

Intriguing Astronomical Scientific Information

Hubble Telescope

- The Hubble space telescope (HST)'s solar panels allow it to store the power of 20 car batteries.
- The HST is 43 ½ feet long and weighs 24,500 pounds.
- The HST is equipped with three fine guidance sensors to assist with the direction of its pointing system.
- The HST was put into orbit around planet earth in the year 1989. It orbits our planet once every 97 minutes.
- Our solar system consists of all of the objects that orbit our home star, the sun. The distance the suns gravity pulls on things is, "its total diameter is around two light years across" (Sparrow 20).
- One astronomical unit (AU), is the earth's average distance from the sun throughout one year (93 million miles). "Mercury's eccentric orbit varies between 0.31 and 0.47 AU from the sun and the planet completes one circuit every 88 days" (Sparrow 21). The planet Venus orbits the sun in nearly a perfect circle. The year on Venus last 225 days and its circular orbit is at 0.73 AU. "The orbit of Mars is also noticeably

elliptical the red planet orbits between 1.38 and 1.67 AU from the sun, in 667 days" (Sparrow 21).

- Mercury's orbit around the sun is so tight that the scientists who control the (HST) are afraid that the planets 88 day orbit does not venture far enough from the sun to risk damaging the (HST).

Intriguing Astronomical Scientific Information

- The photons our sun emits in every direction are pulled inward toward our magnetic poles. This causes the wavelength of light to have a shimmering iridescent greenish hue. Other colors that the light is often times accompanied by are red and violet. They appear to dance in the sky. Another name for this and more commonly known in America is the Northern Lights. These lights intensify when our sun releases solar flares. The scientific name for the Northern Lights is the Aurora Borealis. These lights occur in a layer of the earth's atmosphere called the thermosphere. The thermosphere is a level of atmosphere higher than the hydrosphere (where our weather takes place).
- Venus's surface was mapped using radar beams because it has such a dense atmosphere. One of the space probes that cast these radar beams was on the Magellan mission in the early 1990's.

- Mars has nearly a transparent atmosphere that gives us a clear view of its surface. "The planet spins on its axis in 24:37:00, so it has fairly earth like days, and its axis is tilted at 25° from upright, so it also has an earth like pattern of seasons" (Sparrow 28). Mars has two moons, phobos and deimos. Its gravity is believed to have captured them from the asteroid belt. These are both small though. For example phobos is only 13 miles wide. Since we can see the ground on Mars clearly "spread across an area about two kilometers (1.2 miles) across, the sedimentary rocks resemble those formed under water on Earth , and are strong evidence that Mars once had oceans of its own" (Sparrow 35).
- The exterior of the four outermost planets of our solar system are composed of a gaseous material. Closer to the centers of Jupiter and Saturn are mostly liquid surrounding a rock core the size of planet Earth. Uranus and Neptune have colder interiors which makes an icy slush that surrounds the rocky core.
- When these four planets formed not all the material was captured inside the planet. This left over debris was sent into orbit of each planet. Parts of this system of satellites in some

Intriguing Astronomical Scientific Information

cases formed as moons. There are two different types of moon "in general the moons of the outer solar system fall into two categories, regular and irregular satellites"(Sparrow 40). A regular moon revolves around its planet in the same direction as its planets rotation. It has a circular orbit around planet, that matches the planets rotation. Irregular moons have much bigger elliptical orbits around the planet and it does not coincide with its planets rotation.

- There are many other frozen pieces of rock in the Keiper belt located past Neptune. "Which incorporates dwarf planets, such as Pluto and Eris" (Sparrow 41).

- Jupiter's orbit around the sun varies between 4.95 and 5.46 AU. This takes precisely 11.85 years here on earth to complete one year on Jupiter. Saturn's orbit is between 9.0 and 10.1 au and this takes it 29.5 years on earth. Uranus orbits the Sun at 18.4 and 20.1 AU which a last 84.3 earth years . Neptune makes an elipse around the Sun that varies between 29.8 and 30.4 AU. This takes the planet 164.8 earth years to make one Neptunian year.

- The planet Jupiter is so immense that all seven other planets could fit inside it with room to spare. Jupiter is mainly composed of hydrogen and helium with a variety of other chemical components. Many involving sulfer for which provides its brown orange color to its clouds.
- Inside the planet Jupiter 600 miles deep the pressure is so great that it splits the atoms of hydrogen gas. This creates a sea of electrically charged ions that is like liquid metallic hydrogen .
- Jupiter has three small rings that orbit around it and its gravity hosts a huge family of moons all in orbit.
- In contrast to its huge size the day on Jupiter is very short. Jupiter rotates once every 10 hours, this fast rotation creates a bulge at its equator.
- The Galileo mission in 1995 was an orbiter spacecraft. It was sent to Jupiter equipped with a smaller space probe. Upon deploying its probe into Jupiter's atmosphere, the space probe spent eight years taking pictures and measurements of the planet and its moons.
- Jupiter has 63 moons. Although many of them are just asteroids and comets captured by its powerful gravity. "Close

Intriguing Astronomical Scientific Information

to the planet lie eight that seemed to have formed from the nebula left behind after the birth of the planet itself" (Sparrow 48).

- Four of the moons are called the inner moons, due to their proximity to the planet. The inner moons are small and made of rock. The 4 outer moons are "huge worlds-Lo, Europa, Ganymede and Calisto, each of which has the complexity of a planet in its own right" (Sparrow 48).
- Galileo discovered them through his own invention the telescope in 1609. This was "important proof that not everything in the universe followed a circular orbit around earth" (Sparrow 48).
- Europa is believed to contain a global ocean of liquid water. This is within the crust tens of kilometers deep. The waters are thought to be kept warm by sea floor volcanoes.
- Saturn is the sixth planet from the sun and was the most distant planet known to exist until the invention of the telescope. Saturn is the second largest planet in our solar system. "Saturn is a gas giant dominated by the light elements hydrogen and

helium, which compress into a deep liquid ocean beneath the surface" (Sparrow 52). A day on Saturn lasts only 10.6 hours.

- Like Jupiter's great red spot (which is the most powerful hurricane in our solar system) on Saturn "there is a predictable weather system called the great white spot, which appears near the equator once every 29 years, shortly after Saturn's northern summer solstice" (Sparrow 52). This storm lasts only for a couple of weeks. Since there is no solid ground on Saturn, the storm expands itself completely around the planet. In doing this it then merges with other cloud bands that Saturn has and dissipates.

- During Saturn's summer solstice the planet is tilted showing us it's North Pole. During Saturn's winter solstice, the planet is tilted showing us it's South Pole. "The reason Saturn's rings appear so spectacular is because of their chemical composition. They are mostly made of highly reflective water ice" (Sparrow).

- Saturn is the planet of our solar system with the most debris surrounding it. Not only does this debris assemble its rings that we see but, it exists elsewhere around the planet. "Saturn is accompanied in space by at least 62 moons" (Sparrow 58).

Intriguing Astronomical Scientific Information

Most of the moons are just asteroids or comets that Saturn captured with its gravity. Saturn has at least 24 regular satellites also known as shepherd moons that formed alongside the planet.

- The next planet we come to on our journey is the planet Uranus. It was "discovered by William Herschel in 1781" (Sparrow 60). This was the first planet that was discovered with the help of a telescope. "Astronomers only had the most basic facts an estimate of its mass, diameter and the fact that it had five moons in orbit around it" (Sparrow 60).

- "In 1977, when studying the eclipse of a star that Uranus happens to pass in front of, they discovered that the planet had a series of narrow rings" (Sparrow 60). The discovery of these rings helped to explain some secrets about Uranus. The rings encircled the planet from the North Pole to the South Pole.

- The position of its rings clued us in on its odd rotation. Its moons also rotate around the planet in an alignment that coincides with its rings. "Equatorial regions, experienced a cycle of day and night more in keeping with the planets 17 hour rotation period" (Sparrow 60).

- Uranus completes one revolution around the sun once every 84.32 to Earth years.
- The surface temperature of Uranus is negative 353 degrees Fahrenheit.
- Uranus is axis is tilted 97.8°. This creates unusual seasons on Uranus because for half the year on the planet "one poll is in permanent darkness, while the other experiences endless sun light" (Sparrow 60).
- The color of Uranus is a blue green color. Our scientists have learned from voyager 7's fly by that "it showed that the blue color was caused by the presence of methane in its atmosphere and it revealed that the dark rings were also made of methane ice" (Sparrow 60).
- The way Uranus's moons and rings orbit the planet, "with an axis tilted at 98° it is literally a world on its side" (Sparrow 60).
- Uranus is the seventh planet from the Sun.
- Neptune is the eighth and the last of our solar system's major planets. "Much of what we know about this cold blue world comes from a single space probe fly by in 1989" (Sparrow 62). This space probe was that of voyager two.

Intriguing Astronomical Scientific Information

- This far away from the Sun even the gaseous elements form as ice. "If Jupiter and Saturn are the solar system's gas giants than Uranus and Neptune are its ice giants" (Sparrow 62).

- Neptune's orbit around the Sun is so immense that one Neptunian year takes 164.79 Earth years. Neptune is a cold ball of water, methane and ammonia.

- The weather on Neptune contains extremely strong winds. In fact "Neptune has the highest wind speeds in the solar system reaching up to 2100 kilometers per hour (1300 miles per hour)" (Sparrow 62). Like the other gas giant planet's Neptune is also surrounded by a system of rings. These rings were difficult to see until 1989 when voyager two was passed the planet. "Voyager photographed Neptune backlit by the Sun, it revealed the truth - Neptune's three rings are clumpy – thicker in some places but extremely tenuous in others" (Sparrow 62).

- <u>Event horizon-</u> Is the edge of a black hole.

- <u>Sagittarius A-Star-</u> The super massive black hole at the center of the Milky Way Galaxy is 4 million times the mass of our Sun. This is only 26 thousand light years from Earth.

- Keck Observatory- is located on top of Mt Molokai in Hawaii. The constant movement of Earth's atmosphere bends light to make stars appear to twinkle.
- Birth of a black hole- Hypergiant stars that go into supernova afterwards continue to collapse on themselves to a point smaller than an atom and then the black hole is born.
- Quasar- Is the brightest light in the Cosmos. They are light that is emitted from a black hole that is super massive. These are formed by the black hole consuming too much matter. Collisions of this matter cause them to release gigantic amounts of energy. A quasar is immense luminous jets of light that spread out hundreds of thousands of light years into space.
- Everything in the Milky Way revolves around Sagittarius A-Star, the black hole at its center. The speed of this is 500,000 mps. In 4 billion years the Milky Way will smash into its neighbor the Andromeda Galaxy at 250,000 mph. they will form a super galaxy called Milkdromeda. When the black holes at the center of the Milky Way and Andromeda merge they will form a monster, the survivors will face brutal reorganization under the new master.

Intriguing Astronomical Scientific Information

- What we can see of the Milky Way is one of its spiral arms that is a few thousand light years closer to the center of the galaxy than we are.
- There are four basic types of galaxies
    - Elliptical- Built of old stars and has no motion
    - Lenticular- Consisting of a bulge and a disk
    - Irregular- Has no real shape at all
    - Spiral- This is the type of galaxy the Milky Way is
- The Milky Way is what is called a barred spiral. It has a bar of stars going through the middle and its arms are sprouting out from this bar.
- The Milky Way has four main spiral arms and from the inside out they are named
    - The Norma arm
    - The Scutum-crux arm
    - The Sagittarius arm
    - The Perseus arm

        Our solar system lies between the Sagittarius arm and the Perseus arm. The Sun is located on what is called the Orion Spur.

- The Pleiades Cluster is in the Milky Way's galactic disk. This is a group of stars that have formed from dust and gas. The Pleiades is the closest cluster of stars to the Earth. Another name for this cluster of stars is the seven sisters. It is a grouping of little stars in the nights sky that resembles a little kite with a tail. It was formed 100 million years ago and will be around for at least twice that long, before the galaxies spiral arms tear it apart.
- On a clear night you can see a nebula with your naked eye. Look to the stars along Orion's belt and where you see the little region that looks a little cloudy or milky, that's a nebula. This area is active with stellar formation, which makes the gases around the stars glow.
- Nebulas are the universes recycling centers. This is where dirt, dust and gases are collected to form new stars. This is fueled by the engine of the universe gravity.
- On September 5, 1977, NASA launched Voyager 1 into outer space. This space probe continues to progress further and further into the universe. It is expected to continue to venture further and further into space until the year 2025. This is when

Intriguing Astronomical Scientific Information

the lifetime on Voyager 1's plutonium battery is expected to expire. At that time it is believed that its radioisotope thermoelectric generators will no longer supply enough electric power to operate any of its scientific instruments" (https://en.wikipedia.org/wiki/voyager_1). It will still continue forward but, only with the momentum it has. In June 2010 Voyager 1 was approximately 116 AU or 10.8 billion miles from the sun. Then in September 2012 Voyager 1 was 121 AU from the sun "Voyager 1 was traveling at a speed of 38124.105 mph" (https://en.wikipedia.org/wiki/voyager_1). Voyager 1 reached a distance of 125 AU from the sun on August 2nd 2013" (https://en.wikipedia.org/wiki/voyager_1). Voyager 1 still hasn't gotten to the Kuiper belt or the even more distant Oort cloud and all their comets, that all revolve around our sun. It is estimated that if all goes as it is expected, Voyager 1 will reach the Oort Cloud in about 300 years and won't pass through it for approximately 30,000 years. The year is 2015 and Voyager 1 continues to send our scientists at NASA information and pictures it is continuing to transmit. Even though Voyager 1 has surpassed the last planet in our solar system Pluto, it hasn't even come close to passing other galaxies in our local group. This is only a minuscule portion of the visible universe.

- The next area on our journey into the universe is the galaxy our sun belongs to, the Milky Way galaxy. Our sun resides in one of the Milky Way's outer spiral arms. We are fortunate to be located where we are because; we are near the edge of our galaxy. This gives us a better view of the universe and all of its galaxies. "Astronomers think our galaxy contains at least 100 billion stars and perhaps as many as 400 billion" (http://huffingtonpost.com). It truly is incredible when you consider that gravity is the single force that has initiated all of this motion in the universe. Centrifugal force keeps everything spinning at a steady pace because there is nothing in space, so there is nothing there to give any friction to alter the motion. There is an estimated 200 billion galaxies in the visible universe. Sure collisions do take place between stars and galaxies but, gravity is always there to pull together the pieces and keep everything spinning in its rotation around a central point. Astronomers estimate that the Milky Way Galaxy rotates once

Intriguing Astronomical Scientific Information

every 220-360 million years. Considering the suns position in the galaxy it makes a full rotation once every 240 million years. Everything in the Milky Way all orbits the super massive black hole at its center called Sagittarius A Star. Everything in the Milky Way is believed to be orbiting the galactic center, at a speed of 220 kilometers per second.

- The next region we will discuss is the Local Galactic Group. This is a group of 36 dwarf galaxies. Along with the Milky Way and the Andromeda galaxy or (M31) as it is referred to in the astrologic community. The Milky Way and the Andromeda are the only large galaxies within this grouping of galaxies. Both of these large galaxies hold several hundred billion stars.

- With this extravagant number of stars in our local galactic group alone, it would seem to be an unfortunate waste if our planet was the only home to intelligent life. There are other planets that revolve around the other stars. We have sent a space probe out into interstellar space. This space probe was suitably named Voyager 1.

- Black holes have the strongest forces in the universe. They are black because light cannot come out of them. Any light that goes in gets swallowed up and just disappears. It is believed to be an incredibly deep hole into the universe itself.

- Michelle Thaller (Astronomer) said that "black holes are cosmic organizers, they bring things together."
- Around the edges of a black hole, past a thick cloud of cold dust on the exterior lies an immense disk of cosmic debris. This debris is rotating at millions of miles per hour. The debris consists of gas, dust, shattered stars, and planets. This edge of the black hole is called the "event horizon". No light is emitted from this spot in space, for the gravity is said to be too great.
- The immense gravity obtained by a black hole is so powerful; it is the center point of a galaxy and everything in it, is in revolution around at least one black hole.
- No one knows what happens inside a black hole. We do know that everything continues to gain speed as it approaches a black hole. This speed increases up to the speed of light but, no one has seen what happens then because it is to dark and any added light cannot escape.
- Space and time itself do not behave the same way that we think once entering a black hole. Just trying to understand this

Intriguing Astronomical Scientific Information

occurrence will help us evolve to understand the nature of reality better.

- A black hole forms when a star at least 10 to 20 times the mass of our sun dies. These stars are called hypergiant stars. When it dies the enormous star begins and continues to collapse on itself, to a point smaller than an atom.

- When a hypergiant star gets so massive it starts to run out of fuel it starts dying, it can no longer hold up its own weight and it collapses. This releases a huge burst of energy so large that the only thing bigger was the big bang. This burst is called a super-nova. The remaining parts of the star, keep collapsing and squeezes itself to a point smaller than an atom. The birth of a black hole has just occurred.

- Astrophysicist Doug Leonard was watching a new supernova that appeared in 2005. After looking at the image of a hyper giant star, the once very bright supernova has gone dark. Doug says "According to our best theories and calculations we have just witnessed the birth of a new black hole."

- There are black holes tens of millions of times the mass of our Sun. Astronomers can locate these by observing objects that orbit black holes so close and at such a high velocity, they generate heat and glow. This is how they know a black hole must be there.

- Super massive black holes emit light that can be seen all the way across the universe. When a black hole swallows too much matter, collisions of this matter become very violent. A quasar then erupts out of its center. The quasar gives off immense, luminous jets of light that spray out hundreds of thousands of light years into space. A quasar is the brightest light in the universe. There are billions of black holes in the universe.

- It is believed by some that super-massive black holes don't just anchor galaxies; they may also build and create them. Black holes do this when they attract a star to its gravity. As it does this its mass only grows bigger. This greater mass only acts as an attractant that pulls other stars in closer. This is how the galaxies that we can see today have formed.

- Black holes were originally a theory that was predicted by Einstein. We know now they control galaxies and power the brightest lights in the universe called quasars.

Intriguing Astronomical Scientific Information

- Near a black hole it is believed that space flows into it like water in a stream.

- Einstein believed that near a black hole, gravity would start stretching time. A second would expand to years or even centuries. Seen from a distance these objects would seem frozen in time. It is believed that time would seem normal from the object approaching the event horizon (the edge of the black hole).

- Even Einstein didn't know what happened across the event horizon inside a black hole. Daniele Faccio an experimental physicist is doing work with lasers and creates miniature black holes here on Earth. He is bound and determined to answer this question for Einstein but, has yet to find the answer. He is confident that he is on the right track and will one day solve the mystery.

- Some think that beyond the event horizon everything that got sucked in collides with each other and then vaporizes in an explosion.

- Others predict what is known as a singularity. This is a place where space and time come to a single point and come to an end.
- There have been more than 200 planets outside our solar system, found rotating around other stars that astrologers have discovered. All of which have been thought to be uninhabitable for life.
- Planets and stars both are orbiting their common center of gravity. In most cases it is a black hole.
- Just like sound waves, light waves shift as the object moves toward or away from you. Light from an object moving towards you will look bluer. Light from an object moving away from you will look redder. This is called the Doppler Effect. Using this effect, measurements are taken and this change in the wave length of light is what allows us to detect planets around other stars. The stars also have a wobble that can be noticed when watching them over a period of time. This wobble is a counter reaction with planets that are orbiting it.
- Jupiter orbits the Sun at a distance of half a billion miles.

Intriguing Astronomical Scientific Information

- Earths distance from the Sun varies by almost 4 million miles throughout a year's period.

- Astronomers have found a planet they call Gleeza 436. This planet is believed to be composed of a mixture of rock and water.

- There are seven objects in the sky that have motion separate from the daily rising in the East and setting in the West of all the stars. These celestial bodies are the Sun, Moon, and five planets, Jupiter, Saturn, Mars, Venus, and Mercury.

- Venus has an atmosphere that is composed of carbon dioxide and has a surface temperature high enough to melt lead. Another commodity on Venus is corrosive sulphuric acid rain. Its atmospheric pressure is 100 times greater than Earths. This planet has one of the most hostile environments in our solar system. The planet Venus has such a slow rotation that one day lasts 243 Earth days. This means that it goes around the sun faster than its spinning. A year on Venus lasts only 224 days.

- Mercury orbits the sun once every 88 days.

- The ecliptic is the path the Sun makes among the other stars over a year's period of time.

- The ecliptic is divided into 12 equal parts based on 12 constellations. They noticed these constellations vaguely resemble animals and they named them accordingly.
- Constellations are completely made up. In fact it is very much like playing connect the dots. The distance of stars will vary if you change the point of reference from which you are viewing them. So the constellations we are seeing in the night sky, are specifically unique only to planet Earth.
- <u>Sagittarius</u>- is a constellation made up of half an archer and half horse. This is where the black hole at the center of the Milky Way is located (Sagittarius A Star).
- In 1980, Astronomer Carl Sagan expressed a theory that states "the atoms within the human body have been made for us by previous generations of stars that have died in a supernova. That is the only way that those atoms are created."
- Asteroids are traveling through the solar system at approximately 40,000 MPH.
- Asteroids vary in size from just yards to some asteroids that are hundreds of miles wide.

Intriguing Astronomical Scientific Information

- The asteroid that hit Russia in 2013 was traveling at 12 miles per second. This asteroid was 60 feet wide and weighted approximately 10,000 tons.
- The asteroid belt is located in between Mars and Jupiter.
- <u>Titan-</u> is one of Saturn's moons and it is the most earth like place in our solar system.
- In 5 billion years our sun will expand into a red giant.
- Most of the stars in the Milky Way are orbited by planets.
- Kepler 186 is a star 500 light years away from Earth.
- Without earth's magnetic field, life would be impossible. This is mainly due to the protection it gives us from the solar wind. The solar wind is created by the photons emitted by our Sun. As UV light hits the Earth's atmosphere, it creates a thin layer of charged gas. The Earth's magnetic field squeezes the gas into a huge donut shaped ring around the planet. This donut ring extends outwards and reflects incoming solar radiation. Earth's magnetic field stretches far into space and is ever changing.
- Jupiter's magnetic field stretches for 3 million miles.
- The Milky Way galaxy is made of about 200 to 400 billion stars.
- The planet Jupiter is 1000 times bigger than Earth.

- The planet Venus is the closest in size to the Earth.
- It would take 1.3 million earths to fit inside the sun.
- In 1994, astronomers watched as Jupiter's immense gravitational pull captured the comet Shoemaker Levy 9. Upon experiencing this gravity the comet broke apart into 21 pieces. They called the broken parts of the comet "The string of pearls". As each of these pieces struck Jupiter it caused a multi-mega ton explosion. The explosions left dark scars on the planet, each the size of Earth.
- Astronomers have discovered that as a comet travels around the sun it forms a tail near the planet Saturn and begins to leave a dust trial marking the path it has taken.
- When a comets icy nucleus encounters the massive amounts of ultraviolet radiation our sun emits, it develops a separate blue tail. This UV radiation forms blue plasma around the icy core of the comet and the solar wind pushes this plasma away from the Sun. Scientists call this second tail the plasma tail.
- Our solar system was born 4.6 billion years ago.
- Comets come from two areas in our solar system. They are the Keiper Belt, which is located just past Neptune. The second

Intriguing Astronomical Scientific Information

place a thousand times farther, almost a light year away lies the Oort Cloud. This is a vast region in outer space where comets are numerous as ants on an ant hill. These comets along with meteors and meteorites collect to form the Oort cloud. It is estimated that this region of outer space is teaming with two trillion icy dirt balls.

- Astronomer Michelle Thaller said that "the reason the Oort Cloud was created when the planets were forming gravity kicked out lots of material very, very far away that was collected into the Oort Cloud." She also mentions how some material was completely thrown out of the solar system entirely. Some comets travel in such a big ellipse they venture 1 to 2 light years from the Sun.

- In our solar system just past Saturn astronomers discovered an enormous rocky object over 100 miles across. Those who discovered this phenomenon watched as it orbited towards the sun and formed a tail. Initially they thought that is another planet but after observing this they called it a planet and a comet. They called this first of its kind a Cosmic Bodies Kyron.

- Geologist Lydia Hallis found water that existed during the birth of our planet. She found this water on a volcanic island called Staffa in Scotland. The old water was found in the most unlikely of places. It was found hiding inside basalt rocks. These rocks obtained this water when our planet was formed. Some of this water still exists today. It was found by breaking the basalt open. This revealed a crystalized portion that was taken and analyzed in a lab. Although the crystalized water in the basalt didn't match the water that's in today's oceans, this holds out hope for it to match the water inside a comet.

- It is believed that the interior of a comet has frozen ammonia, methane and carbon dioxide. These can combine to form amino acids, which are the building blocks of life.

- There are currently hundreds of extra solar planets that we have found in the universe.

- In 2015, solar probe plus flew directly into the Suns 2 million degree corona. This space probe answered questions as to how the Sun makes life on earth possible.

- Here are a few simple steps that will help you locate the Andromeda Galaxy in the night's sky. First locate the large

Intriguing Astronomical Scientific Information

square in the constellation Pegasus. It's easy to find because it is one of the largest geometrical shapes in the night's sky. Next find the constellation Cassiopeia. This constellation looks like a slightly misshapen W in the sky. The Andromeda is located in between these two constellations. Using these two constellations make an imaginary line up from the upper left corner star in the Pegasus Square. This line should extend to the star that makes the first point downward in the constellation Cassiopeia. The Andromeda is located just slightly closer to the square of Pegasus. If you look midway between the two stars it will appear below the line you just made.

- We have numerous space probes that are sampling, measuring, photographing and even sniffing the dust and gases of other worlds.
- In 1977, one of these space probes detected liquid water erupting from one of Saturn's moons called Enceladus. These geysers on Enceladus are erupting liquid water into space, where it then freezes and becomes one of Saturn's rings.

- Enceladus is named after a tribe of giants in Greek Mythology. Enceladus continuously dumps six tons of water per minute into its atmosphere.
- The Kuiper Belt is thought to be left over material from when the solar system was formed. It revolves around the Sun beyond the orbit of Pluto.
- Triton is one of Neptune's moons. This moon is special due to the fact that it orbits Neptune in the opposite direction of its rotation.
- Voyager 2 is the space probe that has traveled furthest from earth. It is currently still orbiting into the outer solar system.
- Tritons density nearly matches Pluto's.
- Pluto and Triton are both objects that originated in the Kuiper Belt.
- A typical day's temperature on Venus is around 860 degrees F.
- Saturn's moon Titan is covered by lakes of methane.
- Extra solar planets are planets that orbit stars other than the Sun. In 1995, in the Pegasus constellation 50 light years away from Earth, astronomers discovered the first true extra solar planet. It's called 51 Pegasi B. It was found by finding star

Intriguing Astronomical Scientific Information

wobble. A big planet like Jupiter has a lot of gravity. This causes a star to move a little to account for the planets orbit around it. This technique is called the Doppler Effect.

- The size of our solar system is 10 billion miles across.
- Dr. Michelle Thaller is one of the astronomers who has provided us with this information. She works for the NASA Goddard Space Flight Center.
- Pluto is about 4 billion miles from the Sun. Mercury has an oval orbit around the Sun. At its furthest point it is 43 million miles from the Sun. At its closest point it is just 28 million miles from the Sun.
- The day time temperature on Mercury is around 801 degrees F. The night time temp on Mercury is -297 degrees F.
- The planets each have a different size elliptical orbit around the sun.
- Venus is the closest planet to Earth.
- Although there are times when Mars is closer than Venus.
- Gravity is the universal force that causes motion in the universe.

- I have used lots of information provided by Professor Michio Kaku, a theoretical physicist at the University of New York to make this book.
- Some supernovas are so powerful they are second to the big bang itself, for energy and sheer power. These are so magnificent they can outshine an entire galaxy of 200 billion stars.
- A black hole is the most extreme object in the universe. At its center the laws of physics break down time comes to an end and gravity is infinite. A black hole is a bottomless pit of gravity caused by the death of a star. It is now believed that there may be as many as 100 million black holes in the Milky Way galaxy alone.
- In 2011, astronomers witnessed a gamma ray burst this was one of the biggest explosions ever recorded. It was a flash of radiation brighter than 100 billion Suns. The burst came from a supermassive black hole.
- Once you enter the event horizon (the edge of a black hole), that is known as the point of no return.
- The speed of light is 670,616,629 MPH.

Intriguing Astronomical Scientific Information

- Galaxies orbit each other just as planets orbit stars. Gravity is the culprit causing all of this motion.
- It has been said by scientists that right now in 2015, it is likely the first pioneers who will live on another world have already been born on Earth. It is though that we need a spare planet because our lives are too precious to put all life on simply one planet.
- A typical daytime temperature on Mars is -20 degrees to -40 degrees, "the combination of cold temperatures ranging from -87 degrees Celsius (-125 degrees F) to -5 degrees Celsius (23 degrees F)" (Sparrow 28).
- Astronomers have designed a sort of robot to explore and run multiple different tests in space. It is called A.T.H.L.E.T.E. which stands for (All Terrain Hex Limbed Extra Terrestrial Explorer). (It has been said that each of its limbs is like its own Swiss army knife.)
- The moon is approximately 240,000 miles from Earth.
- A trip to the moon is in the planning stages. It is to have six astronauts and have them live on the other side of the Moon. This will be a disadvantage for them because they will be out of

radio contact with Earth. If the astronauts can live for a year there on the other side of the Moon, we will feel more confident that we can move out into the solar system and settle on more difficult places such as the planet Mars.

- A trip to Mars is six months long. That is when the Earth and Mars are at their closest alignment.
- In outer space astronauts convert their urine to water. Then they use the water to make more oxygen. So ultimately they end up breathing their own pee.
- The first manned mission to Mars is scheduled to be in the year 2040.
- Once humans have been to Mars, the next step will be to transform the planet. This means to make Mars Earth-like. This will be like making a garden of eden on Mars.
- A comet coming in and striking the Earth will reach a temperature of 10,000 degrees F. If that comet were to strike in the ocean, that high of temperature would evaporate hundreds of square kilometers of water. All of the evaporated stream would coagulate in the atmosphere and fall as rain elsewhere on

Intriguing Astronomical Scientific Information

Earth. It is believed that if this were to happen a global deluge of water would drown the Earth for weeks or months.

- It is known that sometime in the year 535 AD something created the most severe cold period in the last 2000 years.
- 11,000 BC some catastrophic event sent an entire culture of people into a tailspin. Some scientists believe this to be the effects of a comet strike. Although to date no impact crater has been identified. An air burst disintegrates all of the solid portions of a comet that make an impact crater. An air burst is the extreme heat produced when a comet hits Earth's atmosphere. This heat causes the comet to explode. This release of energy is called an air burst.
- Our moon is believed to have been created when a comet or meteor, struck the earth. This event was believed to take place early in the Earths life, while it was still a molten ball of magma. Some believe that most of the remnants from this impact got broken off into space and combined to form the cold ball of rock that we call the moon today.
- Here is an interesting fact about the moon; I think few people take into consideration. Did you know that the same side of the

moon always faces the Earth? The image that man has made into "the man in the moon" is only due to the fact that the Earth acts as a shield and blocks meteors, meteorites, and other space junk from hitting the moon. This became obvious when we saw how scared the other side of the moon is.

- Columbia University's geologist Dallas Abbott does research to find potential cosmic craters on Earth. Bruce Masse had found a group of discoveries that potentially could be the scars left in the Earth from a cosmic impact. Dallas investigated Masse's discovery's in the Indian Ocean. Abbott used satellite pictures of the depth variations of the ocean floor to locate an area she named Berckle Crater. Berckle is a huge depression 1000 miles off the coast of Madagascar, in the Indian Ocean. This hole in the Earth is 18 miles in diameter and sits at a depth of approximately 1300 feet. A comet striking the Earth in the ocean would cause a large tsunami that would create huge chevrons. These are large v-shaped sand dunes that point away from the comet strike. Although not everyone agrees that chevrons are evidence of a tsunami. Dallas Abbott went to Madagascar and found the evidence she needed to prove that

Intriguing Astronomical Scientific Information

the chevrons where formed from a tsunami. She found this proof when she discovered marine fossils that were deposited 200 meters on top of nearly all of these dunes. The type of fossil Abbott found were from the bodies of marine animals that live on or near the bottom of the ocean. This eliminates them from being deposited from normal weather conditions.

- There is also another scar or crater under the ocean surface off the coast of Australia. This is in fact two separate impact zones. We have observed a comet hitting the planet Jupiter. When this comet hit the atmosphere it heated up so much it split into pieces. This is what is believed to have happened millions of years ago here on planet Earth near the coast of Australia.

- The Milky Way galaxy is 100,000 light years in diameter. Light travels 186,000 mps. One single light year is equivalent to about 5,878,000,000,000 miles long. So it is simply mind boggling when you consider that the Milky Way is just one of hundreds of billions of galaxies in the visible universe.

- The Milky Way galaxy is our family of stars that we travel through the universe with. All of the stars are orbiting a common center. This common center is that of the black hole called Sagittarius A Star. A black hole is black due to light not being able to escape its grasp. Our black hole has so much gravity it keeps the 200 to 400 billion stars in the Milky Way galaxy orbiting it.

- We cannot see the center of the Milky Way Galaxy. What we can see of the Milky Way is one of its spiral arms that is closer to the center of the galaxy than us.

- When we look at the Milky Way from our Earth bound position it is like looking at the edge of a coin and we get no sense of the galaxies real shape.

- It takes our Sun and planet 200,000,000 years to make one lap around the Milky Way. That is how large the Milky Way Galaxy is.

- Our Sun is located on one of the outer arms of the Milky Way. Lucky for us this gives us a better view of the universe and all of the countless galaxies out there.

- Light from the majority of the stars not only in our galaxy but, all the stars in the universe are greatly dimmed due to the dust that is everywhere in the universe.

Intriguing Astronomical Scientific Information

- Scientists have calculated that the star Rigel, the star that makes Orion's left foot is 772.96 light years away. This means that it took the light from that star 772.96 years of traveling through outer space at the speed of light to reach planet Earth.

- Asteroids and comets it is believed, may have delivered the building blocks for life to planet Earth.

- The question isn't whether we will be struck by a comet or asteroid in the future but, when and where?

- Asteroids and comets are currently cosmic enemy number one but, at the same time they could deliver humans vital resources that could one day save us from ultimate extinction.

- Approximately 4.5 billion years ago our solar system was formed as celestial rocks that collided to form planets.

- Comets are currently prevalent in the Keiper belt outside of Neptune. The Keiper belt is much larger than the asteroid belt between Mars and Jupiter.

- The Oort cloud is a spherical deep freeze reservoir, the forming sun flung comets to the outer regions of the solar system.

- Scientists have discovered there are what they have named main belt comets inside the asteroid belt. Perhaps during

Earth's formation some of these main belt comets may have delivered H2O to the Earth.

- 100 tons of meteorites and space dust falls to earth every day.
- The gravity of Neptune can pull comets out of the Keiper belt.
- 65,000,000 years ago a cosmic object around 6 miles across struck planet Earth on the eastern coast of Mexico. It actually struck in the water so it sent out tsunami type waves. All of the molten debris from this covered all of North America. Scientists believe that this event extinguished 75% of all life on Earth; the dinosaurs are thought to be one of the casualties.
- Jupiter acts as a planetary shield for planet Earth; its strong gravity has captured comets that could possibly impact Earth.
- There is thought to be over 100 billion galaxies in the visible universe.
- The entire vastness of the universe is difficult to comprehend but, here is a way to consider it. If you were to shrink the size of the sun down to a marble and put it on a sidewalk in downtown Manhattan. The Earth would be a pinprick about 4 feet away, Mars would be 2 feet beyond that. The nearest other star, where we might find intelligent life would be Alpha

Intriguing Astronomical Scientific Information

centauri. This star would be another marble on the sidewalk in downtown Washington, DC 230 miles away.

- 370,000,000 miles from Earth orbiting the gas giant Jupiter is the most likely place for life in our solar system. It is an ice covered moon of Jupiter called Europa. It is believed that the ice on Europa is covering an ocean of liquid water.

- It is believed that our first contact with alien life will probably be with intelligent machines. This is believed to be true because it is easier for a machine to tolerate extreme conditions of outer space. At least every living organism we have seen is to delicate to survive in outer space.

- Red dwarfs are stars that are not much larger than planet Earth. The vast majority of stars in the universe are red dwarfs.

- Wikipedia an online encyclopedia estimates there are 10,000,000 super clusters in the visible universe. The Milky Way galaxy is a member of the local group. There are 56 other galaxies that make the Local Group. The Local Group is merely a part of the Laniakea Supercluster.

- There are over 500,000 asteroids in our solar systems asteroid belt.

- It is 2015 and sometime within the next two years an asteroid, scientists have named Bennu will pass by the earth closer than the moon is to us. NASA has a list of about 1,500 asteroids that could one day threaten earth. Each asteroid is over 300 feet wide and ranked among its size, speed, and chance of impacting Earth. Bennu's close orbit puts it at number two on the list.
- The sun isn't the only thing that affects the orbits of objects in our solar system. The planet Jupiter is almost 90,000 miles in diameter. Within Jupiter's extreme pressure it squeezes a soup like existence of hydrogen, into a type of substance similar to liquid metal Jupiter has a huge mass that pulls on everything around it, especially the asteroids that are close to it.
- Massimiliano Vasile is an aerospace engineer who believes that the heat produced from a laser will be enough to move an asteroid in space.
- The largest asteroid in the asteroid belt is called Sierries and is the size of Texas. Plumes of water vapor erupting from Sierries have been recently discovered. These plumes suggest that there are vast reservoirs of water below. Scientists believe that as much as 25% of this asteroid could be water. That is more

Intriguing Astronomical Scientific Information

fresh water that exists on planet Earth. This water can be used to fuel rockets like a gas station.

- Astrophysicist Hakeem Oluseyi says that "asteroids are a big opportunity for humanity because ultimately we're going to have to leave the planet." Asteroids that come near the Earth contain resources that we could use to expand our exploration of the universe. NASA is planning a voyage to an asteroid that is planned to take off sometime in the 2020's.
- It is believed that there are literally an uncountable number of planets where life could exist in the visible universe. We are still looking in 2017 but, have yet to find one.
- Leading astronomers and Astro biologists have applied the principles of evolution and physics to approximate what types of creatures us humans could expect to find living on alien worlds elsewhere in the universe.
- We currently know that the universe contains billions of galaxies and trillions of stars.
- The habitable zone of a star is the distance that water can exist on the planet of that star in a liquid form. Liquid water is one of the basic elements for life as we know it to begin.

- Planets with a much higher field of gravity than planet Earth would contain a denser atmosphere, making it easier for an organism to fly.
- Search for extra-terrestrial intelligence or S.E.T.I. is an institute that primarily searches for aliens.
- Flying saucer reports from many different people throughout the world, have nearly all reported UFO sightings have that have made maneuvers unattainable with our modern day aircraft. Not only that but, our human bodies could not withstand such drastic changes in inertia.
- UFO reports all have this in common; it is that the object appears to move without making any sound.
- Alpha Centauri is the closest star to our sun and is 4 ½ light years away. Our farthest distance us humans have sent a space probe is voyager one. It was launched in 1977 and has been flying for 34 years. Traveling at a speed of 30,000 miles per hour and is only now starting to leave our solar system.
- The speed of light is 186,000 miles per second.
- In theory using fusion power in space flight would allow you to travel to another star mine the helium that is there and use it to

Intriguing Astronomical Scientific Information

fuel your fusion engine. The Max speed of a fusion powered spacecraft is believed to be 28,200 miles per second, at that speed it could reach Alpha Centauri in 35 years. At that speed it would be difficult for our bodies to function properly.

- Gamma rays are the highest known energy source in the universe; these potentially are the key to interstellar space travel.
- In theory, a thinking machine or artificial intelligence is believed to keep on getting smarter. So if someone were to create one of these machines 100 years from now it would be smarter than all of the humans that have ever lived.
- Our sun is one of billions of stars in the Milky Way galaxy. The Milky Way is just one of hundreds of billions possibly a trillion galaxies in the universe.
- A galaxy is a collection of a hundred billion or so of stars.
- When we look at some of the most distant galaxies, the light we are seeing was emitted from those stars before the Earth was even in existence. The Earth is 4.6 billion years old and some of these galaxies are believed to be 13 billion years old.
- The Andromeda galaxy is about two million light years away.

- A quasar is the active galactic nuclei that come out of a black hole at the center of a galaxy. They are very powerful engines that produce a tremendous amount of light.
- The brightest known quasar is about a trillion times as bright as our sun.
- The Andromeda galaxy has over twice as many stars as the Milky Way Galaxy.
- Andromeda galaxy has a double nucleus two huge clumps of stars at its center. That means it has two black holes at its center.
- Gravity is the force that is pulling the Andromeda and Milky Way galaxies together. They are actually approaching each other at an astounding rate of several hundred kilometers every second.
- The birth of a white dwarf takes place when the gas at the center of a star like the Sun is heated by high energy radiation. This takes place at the end of the stars life. Prior to this occurrence, the star emits an expanding shell of gas that is called a planetary nebula. This gas appears with brilliant colors, "the blue green color is due to glowing oxygen atoms and the

Intriguing Astronomical Scientific Information

red hues are produced by hydrogen and nitrogen" (Schilling 86).

- "There is a helix nebula in the constellation Aquarius. It is releasing thousands of strange elements of gas with a head and a tail making them look like tadpoles, this is coming out of the largest planetary nebula in the night sky"(Schilling 91). In star terms planetary nebula have a very short lifetime, they will only be visible for an estimated 20,000 years at most. A nebula is a common occurrence that happens when a star like our sun dies.

- Here is an example of size comparison of a white dwarf, "white dwarfs can be more massive than the sun, but they are about the same size as our Earth" (Schilling 94). At the end of our Suns life it will become a white dwarf. First, it will swell into a red giant and expel its gas mantle out into space as a planetary nebula.

- At the end of a massive stars life, it ends itself when it forms stable atomic nuclei in its core. This causes the star to stop making any more fusion reactions. Here is what happens next... If the star is massive enough, "the star collapses under its own weight and with such violence that it then blows itself apart in an unimaginably powerful explosion. Which is visible to the naked eye even at distances of hundreds of thousands of light years away" (Schilling 95).

- One of the Milky Way's neighbors is 160,000 light years away. When a supernova happens it produces so much light that to us from our distant position here on earth, it seems as if there is a new star added to the sky. A supernova can be visible for many months, then it vanishes from sight and a black hole is left in its place.

- The stars have always fascinated us humans. Danish astronomer Tycho Brahe published an essay in 1573, and since then the exploding star has been known as Tycho's supernova. Tycho's supernova is currently 24 light years in diameter.

- Scientists have discovered a pulsar star that is accompanied by a neutron star. These two stars are currently in orbit with one another at an average distance of a few 1,000,000 kilometers, "In about 300,000,000 years, the two stars will collide and merge to become a black hole" (Schilling 108).

- On average are closest planetary neighbor Mars, sits 142,000,000 miles from earth. Due to the planets orbit around

Intriguing Astronomical Scientific Information

the sun, there are times when Venus is closer than Mars to the Earth.
- Mars is the fourth planet from the sun and the average temperature is negative 80° F.
- In August 2012 NASA sent a machine two scout out the conditions on Mars. It was a vehicle they named the Mars curiosity rover. It is a one ton robot that is the size of an suv. The price tag on this suv was around 2.5 billion dollars. At the time of 12/4/14 the rover had successfully driven over 5 miles of the martian surface.
- Mars has ice caps at each of its poles. That is frozen water believed to be over a mile deep into the crust on Mars. If these ice caps melted away they would cover the planet with over 60 feet of water. Theoretical physicist Sylvester James Gates JR. has stated the fact that water is a precursor for life itself, if there's water there's a possibility of life. So maybe martians are not science fiction.
- Just below ground level on Mars lies an area called the cryosphere. The cryosphere is six miles thick and is thought to hold 50 times more water than the ice caps themselves. Life on

Earth has been found in mines 3500 feet below the surface. All that I'm saying is that there potentially already could be life on Mars. The environment in our salt mines here on Earth mirror the conditions that exist on Mars.

- It is believed that Mars quakes are common on Mars. They are thought to be equivalent to a magnitude seven earthquake here at home.
- Olympus Mons is a volcano on Mars that is the biggest volcano in our solar system. The land mass is the size of Arizona and is three times taller than Mt. Everest.
- The molecules that make up the atoms in our human bodies have been distributed from stars that have died in a supernova explosion. This explosion distributed the materials that were used to make our bodies.
- The closest star to the sun it is 4.367 light years away, it is the star Alpha Centauri.
- When we look at the stars on a clear night, you could only see around 2 to 3 thousand stars. Now if our sun was located near the center of our galaxy there would be a million stars in the night sky as bright as the brightest star that we can see in our

Intriguing Astronomical Scientific Information

night sky. The light from all of the stars would light up the sky so much there would be no difference between day and night.

- Michelle Thaller believes that the Milky Way's spiral arms are evidence that other small galaxies have collided with our Milky Way, to give us the four spiral arms that we have today.
- Our sun is currently in stage 4 of its life. Most of a stars life is spent in stage 4 as a main sequence star, about 10 billion years.
- The center of the Milky Way is crowded with stars that have reached the red giant stage of their life. These red giant stars are hundreds of times bigger than our sun.
- When the core of a red giant star collapses, it causes a supernova explosion. If the stars mass is large enough, after the explosion what is left is a black hole. Around a black hole exists a stellar nursery. The stars that are closest therefore moving fastest are the young stars that have very recently formed. Once a star goes through a supernova what remains is a tiny, very dense neutron star.
- Most of the stars you see in the night sky are a pair of stars known as a binary star. These stars orbit close to one another attached to each other by the engine of the universe gravity.

- The final stage of a super massive stars life is when its core contracts to become a black hole.
- Our sun has been around the Milky Way 18 times since the formation of the galaxy.
- A star that is between 1.5 and 3 solar masses contracts to become a neutron star. If the stars core is greater than 3 solar masses, the core contracts to become a black hole.
- The expansion of the universe only applies to objects that are not bound together by gravity. For example, the Earth is not getting any closer or further than normal from the sun as time goes on.
- The Andromeda and Milky Way galaxies will collide in the next few 1,000,000,000 years. Most likely no stars will actually collide; the space between them is just too great. Some stars will get thrown into the black holes in the middle of each galaxy. Some things will get ripped off and thrown out into space; it will be a very dramatic collision.
- Elsewhere in the universe almost all of the alien galaxies that we see are moving away from us. The more distant a galaxy is from us the faster it is moving away.

Intriguing Astronomical Scientific Information

- The Milky Way is a member of the Local Group. The Local Group is a small cluster of about three dozen galaxies. (The Milky Way and Andromeda galaxies are the two dominant galaxies in the Local Group. Most of the other galaxies are small dwarf galaxies).
- Every galaxy has smaller galaxies that orbit around them. All of the bigger galaxies feel the influence of each other's mass in orbit around each other. This is around 30 or so galaxies that are all orbiting around in the universe. This is a tremendous amount of motion when you consider that there are billions of galaxies in the visible universe.
- The black hole at the center of the Milky Way is four million times the mass of our sun. The edge of it is called the event horizon, which is the point of no return. Not even light can escape this immense gravity, it is a supermassive black hole.
- Less than 0.5 to 1% of our universe is believed to be made up of observable matter.
- Dark matter exists in the universe. It contains a mysterious type of gravity that gravitationally pulls on things. Dark energy is causing the universe to expand, as each second passes. It is

causing the speed of this expansion to increase to a faster and faster speed with time. Both of these units are difficult to describe because they are invisible.

- It is estimated that in perhaps a billion or trillion more years the fate of the universe and the impact its black holes, dark matter and dark energy have on mankind will finally be known. It does not look very promising for mankind right now though.

- Here is a look at the universe, from our position in the solar system. Since gravity is the culprit that has caused and continues to cause motion in the universe, it is the force that gives the universe order. In our solar system our planet Earth is the third planet from the sun. It goes in order Mercury, Venus, Earth, Mars, the asteroid belt, Jupiter, Saturn, Neptune, and Pluto. Beyond Pluto lies the Keiper belt. It has been discussed that Pluto may have been a meteor that has drifted out of the Keiper belt and has been captured by the gravity of our sun. The Keiper belt is a vast region of comets, liquid, frozen gases, ammonia's and rock that surrounds our solar system. The fact is, the Keiper belt is 20 times as wide as the asteroid belt and is 20 to 200 times more massive. The Keiper

Intriguing Astronomical Scientific Information

belt was named after a Dutch American astronomer Gerald Keiper. "Even though Gerald did not discover the belt he did however discover the first Keiper belt object (KBO) since Pluto" (keiperbeltwhen.com). The Keiper belt is home to a number of comets that have a 200 year revolution cycle around the sun. Beyond the Keiper belt lies the Oort cloud. This is yet another vast region of even more massive pieces of frozen water, gases, ammonia's, rock and is home to many comets that all revolve around our sun. Here is a little fact about comets, they only have a noticeable tail when approaching or receding the sun. Their tails are remnants that are melted off their frozen bodies and are pushed away from the photons emitted by the sun. We humans see the photons as light. Since the moving photons are what's making a tail of the comet, the comet's tail always point away from the sun whether it is coming or going. Comets from the Oort cloud have a revolution around the sun that is so big it only passes the sun once every 10,000 years.

- Proxima Centauri is the nearest star to the Earth, besides the sun of course. It sits 4.2 light years from Earth. So when you consider precisely how long one light year is this star is 24,687,600,000,000 miles away from the Earth. This Sun and Proxima Centauri both reside in the Milky Way galaxy. The Milky Way galaxy contains somewhere in the vicinity of 100 to 400 billion stars. The Milky Way is just one galaxy among about 200 billion galaxies in the observable universe. Since we are inside of the Milky Way, it is difficult for us to know its exact shape. We have learned much of what we now know about our galaxy from looking at other galaxies. This is similar to how one would never know what their eye color is if they never saw their own reflection. The Milky Way is what is called a barred spiral galaxy. The Milky Way is 100,000 light years in diameter. The Milky Way is somewhere between 6 thousand and 12 thousand light years thick. As you gaze into the sky on a clear night and your eyes scan over what we can see of the Milky Way you can only see of about 0.0000025% of the galaxies hundreds of billions of stars. The galaxy has several types of energy in it, these include the visible light that we can see, infrared light, radio waves, gamma rays, dark matter and x-rays. The Milky Way along with about 40 other galaxies

Intriguing Astronomical Scientific Information

make up the Local Group. The Local Group belongs to an even larger group of galaxies called the Local Super Cluster. This Super Cluster of galaxies is an astonishing 100 million light years across. All of this is just a piece of the observable universe.

- It takes precisely 299,792,458 meters per second for light to move. When you consider our average distance from the sun, it takes the suns light particles 8 minutes to reach planet Earth. It takes the light from our nearest neighboring star, Proxima centauri (4.2 light years away) a little over four years to reach us.

- Somewhere around 20% of all the galaxies have invisible spiral arms that extend beyond the visible ones. These arms contain young stars that burn so hot that they give off invisible ultraviolet light.

- Astronomers have created a unit of distance to help to measure the vast distances within our solar system. This unit is called an astronomical unit. One astronomical unit is the average distance between the earth and the sun, which is about 93,000,000 miles.

- "Using the standard scale light travels at 186,000 miles per second, that is roughly 6,000,000,000,000 miles per year" (notes from has Our Night Sky DVD by Professor Edward M Murphy PHD University of Virginia).
- If you were to look at the constellation Orion in the southern hemisphere south of the equator, he would appear to be standing on his head upside down.
- Sirius is the brightest star in the night sky.
- The last two and stars in the cup of the big dipper are aligned to point in the general direction of the north star, Polaris. The whole sky rotates around Polaris throughout one year period of time. The motion of this rotation allows for the stars to rise in the East and set in the West.
- Where there isn't any light pollution like light from cities, the moon, cars, homes, and other forms of manmade light, you can see stars, constellations, planets and nebula in the night sky. Nebulas are clouds of interstellar dust and gas.
- Scientists caught the sound waves created by the merging of two black holes. This was evidence that black holes truly exist. Black holes have so much attraction to everything in the

Intriguing Astronomical Scientific Information

universe, that they swallow everything that comes near them even light. When the two black holes combined together it sounded like static, followed by a drop of water dripping into a bucket of water.

- Our sun is pretty normal when compared with the majority of other stars. Our sun is approximately 1.5 million km across. It has a surface temperature of roughly 5000-6000° C. Our sun started burning about five billion years ago, it is believed to have about five more billion years left in its life. The majority of stars are believed to have a 10 billion year lifespan. Of course this number is merely an estimate or an approximation because the life of a star is so tremendously long compared to a human's life. In fact, a stars life overwhelmingly surpasses the existence of all mankind, that a fair comparison would be like comparing the entire planet Earth with a single atom.

- There are particles of gas and dust just floating out in deep space, far from any existing star. Each of these particles contain a minute amount of gravity. This gravity that each particle has causes the particles to attract each other and to form immense clusters. The nearest of these clusters to Earth is 1500 light years away. This cluster can faintly be seen along the constellation of Orion's belt. Although the entire cluster is as extensive as the whole constellation, the clustering of enough material near Orion's belt forms glowing gas, nebula and newborn stars. The cloud of gas and dust is believed to be so extensive that it contains enough material to form many thousands of stars.

- The Rho Ophiuchus cloud is named this due to its proximity to a star in the constellation Ophiuchus. "This is one of the closest star forming regions to Earth, at a distance of about 450 light years" (Schilling 29). Let me put this distance into perspective for you. Since the speed of light is 670,616,629 MPH and a year has 8760 hours. So in order to find it out how many miles from Earth this star forming region is you would just multiply the speed of light by the number of hours in one year. Then you would take that product and multiply it by 450 to see how far it is to the closest star forming region is to Earth. You can do the math if you want but, I'll just say it's an exponentially big number!

- The Earth is orbiting the sun at about 66,000 miles per hour.

Intriguing Astronomical Scientific Information

- The universe contains over 100 billion galaxies, with each galaxy containing 50 billion to some having over a trillion stars.
- The planet Jupiter has a mass of four octillion pounds. Most people have no idea what an octillion pound is but, this would be a 4 followed by 27 zeros. Jupiter is 300 times more massive than the Earth. The sun is about 1000 times more massive than Jupiter.
- The most massive star that we have found to be in existence in our galaxy orbits around the Milky Way, in the large Melangellic cloud. This star is in a nebula within these clouds, called the Teranchella nebula. This massive star is called R136A. It is a young star that is around one million years old, its surface temperature is 70,000 degrees. R136A is 100 to 300 times the mass of our sun.
- The sun is 870,000 miles across. That is 109 times as wide as the Earth. Over one million Earths could fit into the volume of the sun.
- Betelgeuse is a super enormous star in the constellation Orion. That star is less than half the size of VY Canis Majoris, the

largest star that we have found in the universe. Betelgeuse is 1000 times the size of the sun.

- Since there appears to be enough dark molecular clouds in our galaxy alone, this doesn't even consider the material in the billions of other galaxies in the visible universe. "These clouds usually contain sufficient gas and dust to form several 1000 stars" (Schilling 40) So the scenario I mentioned where the universe ends with everything that is will be stuck in an eternal deep freeze, will have to wait for all of the star forming regions in the universe to be exhausted and stop making new stars. As long as gravity continues to pull on atoms and cause motion in the universe, it will continue to ignite those combustible atoms and combine them to fuel the stellar nurseries. That is what generates the birth of new stars. You may wonder just how long it takes for a star to be born. It is stated here in the Govert Schillings book Deep Space "So it is not surprising that the birth of a star also takes much longer, easily at least 100,000 years" (Schilling 48). Given that information the universe will have enough material to keep the star making cycle going for many trillions of years.

Intriguing Astronomical Scientific Information

- As gravity continues to pull on every atom in the universe, it will continue to make new stars from the sporadic clouds of gas and dust in the universe. As these new stars are condensed by the gravity that every atom contains in the universe "not all the material ends up in the star" (Schilling 49). This material will be captured by the newly formed stars gravity. It will then continue to orbit the star, rotating at its own individual rotation. This is how stars give birth to planets, "with a little luck the material in the disk will clump into one or more planets" (Schilling 49). As the material that is left over when every star is born, is cast into a rotating disk which becomes a protoplanetary disk.

- Now that we are gaining knowledge of what commonly transpires when the birth of a star takes place. We have focused our attention as we look at other stars we know that since it has happened with our own star the sun; other stars must have planets orbiting them too. "Dozens of protoplanetary disks have been discovered in the Carina nebula, a giant star forming region at a distance of about 7000 light years away" (Schilling 49).

- With all of the trillions of stars that exist in the visible universe it would seem entirely unlikely for planet Earth to be the only planet in position the exact distance from its star to have liquid H2O on its surface. Since liquid H2O is a precursor for life as we know it to exist.
- If you are like me, you've had the feeling of falling when you are in a seated position. The "great attractor", "a colossal concentration of material for to which the Virgo super cluster including our own Milky Way is falling at a velocity of 600 kilometers per second" (Schilling 175). Maybe that motion is the reason for us feeling like that.
- The big bang happened 13.77 billion years ago.
- This is a easy way to consider the expansion of the universe, "the galaxies are light spots on the surface of a party balloon when the balloon is inflated the spots move further apart without actually moving across the surface of the balloon" (Schilling 185). This expansion of the universe is trumped by gravity, so the Earth will not move any further from the sun in any given year.

Intriguing Astronomical Scientific Information

- Just pretended it is the year 2000, the light coming from a star cluster 6400 light years away left those stars in the year 4400 BC. As you look into the night sky, the light you are seeing from the stars is old light.

- "The existence of dark matter has so far been proven only indirectly on the basis of its gravitational effects" (Schilling 194). Nobody knows how we could even get a sample of dark matter to test. We just know that something is out there in the universe and calling it dark matter seems to offer the best explanation.

STARS

- In the constellation Orion, the brightest star in the right shoulder is this star named Betelgeuse. The bright star that is his left foot is this star named Rigel.

- Blue stars are the hottest, and red stars are the coolest. The temperature of a star is measured in Kelvin. For example 500 Kelvin is equal to 440.33° F. Blue stars are ranging from at least 3,000 Kelvin and red stars are about 3,000.

- The temperature of a star is divided into seven classes, O stars are at least 30,000 degrees, B stars are 20,000, A stars are 10,000, E stars are 7000 Kelvin, G stars are 6000 Kelvin, K stars are 4000 Kelvin, and the M stars are 3,000 Kelvin. (The star Rigel is a B star and Betelgeuse is a M star.)

- When we look at stars the distance between them is millions or billions of miles apart. When talking about distance in our solar system the measure of distance that is used is called an astronomical unit AU. One AU and is the average distance between the Earth and the sun throughout one year or 93,000,000 miles. The speed of light is 186,000 miles per

Intriguing Astronomical Scientific Information

second. It takes light reflected from the moon 1.25 seconds to reach Earth. So the moon is 1.25 light seconds away. It takes light from the sun 8 minutes and 20 seconds to reach the Earth. Jupiter is 43 light minutes from Earth and Pluto is five light hours and 20 light minutes away.

- Though voyager one spacecraft is the most distant spacecraft that we have launched and it was 16 light hours from planet Earth in 2014.

- A light year is the distance light travels in one year. Other than the sun, Alpha Centauri which is 4.4 light years from earth and is the closest star to Earth. So it takes its light waves 4.4 years to reach us.

- The typical star you can see at night is many tens or hundreds of light years away. So as far as a star is from us, it has taken it's a light that long to reach the Earth. So in essence the light waves you are looking at are tens or hundreds of years old.

- The stars have fascinated mankind from the beginning of our existence. The first person that has made a telescope using a convex mirror was Isaac Newton in 1688. "He devised the first reflecting telescope in which the objective is a convex mirror"

(Schilling 52). It was really ingenious of him to think of using a mirror in the telescope, that had a shape like the exterior of a sphere. This creates kind of a magnifying glass out of the eyepiece. Mankind has continued to improve upon Isaacs design, for the 328 years since he invented it. I am certain that we will keep making improvements that will surpass our current telescopes because that is just the nature of our species, to improve upon what we have. The number of extra solar planets has and will continue to increase as we study the cosmos.

- Scientists have learned that the universe can be compared with a recycling plant. When a star dies, it ends its life by exploding in a supernova explosion. This huge explosion blows a large portion of the stars matter back out into space. This is how the universes recycling center works. "From the stellar gas enriched with newly formed heavy elements, new stars and planets can be formed" (Schilling 57).

- The star Polaris is the north star. The position Polaris is in the universe gives us the illusion that all of the other stars in the universe are in rotation around it. When in fact it is only the

Intriguing Astronomical Scientific Information

Earth that is rotating, while it is going around the sun. Making all of the other stars in the celestial sphere appear to rise in the east and set the west. Polaris is a yellow giant star and it is around 50 times the size of the sun.

- The formation of a star begins when an interstellar cloud of gas, that is outer space, very slowly begins contracting into a more and more dense ball of gas.
- <u>Main sequence star</u>- fuses hydrogen into helium. This type of star remains in this state for a long time. The sun is a main sequence star.
- <u>Red giant star-</u> the stars core grows hot enough for helium fusion to begin. The interior of the star reaches 3,000,000° K, while its exterior is only 3000 to 4000 Kelvin.
- <u>Yellow giant star-</u>its core allows it to reach helium fusion rapidly allowing expansion.
- <u>Second stage red giant star-</u> exhausts its helium in its core and shrinks while the exterior expands.
- <u>Planetary nebula-</u> Is a luminous cloud of gas that surrounds a dying star.

- White dwarf star-slowly cools until it collapses, and blasts itself apart in a supernova explosion. What is left after this catastrophic explosion is either a neutron star or a black hole.
- There are a lot of types of stars in the universe, "red dwarfs are by far the most common star in the cosmos" (Schilling 64). Our own galaxy contains several hundred billion red dwarfs. One reason there are so many red dwarfs is because they're not a very hot type of star, they burn for so long some believe them to have eternal life.
- Here is basically how astronomers figure out how far a star is from our planet. They use a theory discovered by American astronomer Henrietta Leavitt. She states "it's quite simple they measure the pulsation of the star, then by applying Leavitt's law, calculate the average luminosity of the star" (Schilling 67). Mathematically using its observed brightness (luminosity), in comparison against the night sky one can calculate how far away the star is from the Earth.
- We know that other stars have planets orbiting around them because using our computers we can notice slight variations in the luminosity or brightness of a star. This tells us that the

Intriguing Astronomical Scientific Information

reason a star temporarily goes dim or loses its brightness is because an orbiting planet intercepts a small fraction of the stars light, as the planets orbit between the star and the Earth.

- As we study stars that are close to us, for example the star Fomalhaut is one of the brightest stars that can be seen in the night sky. This is due to the fact that it is only 25 light years from the Earth and it radiates as much as about 17 times as much a light as the sun. This star is in the constellation Pisces Austrinus, "which is also known as the southern fish" (Schilling 50). In orbit the Hubble space telescope discovered a protoplanetary disc in 2008. One big discovery is what is thought to be a Jupiter size planet that is only seen as a spec but, could clearly be seen with the help of this powerful telescope. Others assume that "Fomalhaut may be a relatively compact dust cloud with an embryonic planet at its center" (Schilling 50). If we can barely see a Jupiter size planet orbiting another star, the opportunity that another earth like planet exists somewhere in the cosmos is so mind bogglingly likely. When you take into consideration how many trillions of stars that there are in the observable universe, this should seem obvious.

- There is an inhabitable zone around a star. Inhabitable meaning an area where H_2O could exist on it without being too close to the star that it would be boiled away, or too far away from a star in having any H_2O will be frozen solid. This is significant because every form of life on Earth requires liquid H_2O to survive.

Intriguing Astronomical Scientific Information

THE MILKY WAY GALAXY

- The Milky Way galaxy contains a few 100 billion stars other than the sun. We set 27,000 light years from Sagittarius A-Star, the black hole at the Milky Way's center. There are so many dust clouds in outer space that absorb so much of the light from the trillions of stars that are in the universe. This is one reason why we can only see a very small portion of the Milky Way.

- When we look into the sky at the Milky Way, it surrounds our planet like the belt. Nearly exactly one-half is above the horizon and the other half is below the horizon.

- There are hundreds of billions of galaxies in the observable universe, that we currently have found to be in existence. Early astronomers thought that they had some idea as to how big the universe was but, "in the 1920s that became clear that the milky way was just one of countless galaxies in the universe" (Schilling 114). So our knowledge about all of the planet's, stars, galaxies and groups of galaxies will only improve as we continue to improve and make adaptations to our telescopes.

- We have found a way to see beyond the dust clouds that cluster the universe. We do this by "using an infrared telescope which can look right through the dust" (Schilling 116).
- When you consider that the speed of light is 670,616,629 MPH. Then you take into consideration that one light year is the distance that light travels through space in a year's period of time. The Milky Way has a diameter of around 100,000 light years. The bulge at the center of the Milky Way is estimated to be 10,000 light years in circumference. Where the majority of the Milky Way itself is in a "disk that is only a few 1000 light years in thickness" (Schilling 117).
- The gas and dust that has made the Milky Way galaxy has been deposited by other stars that have died in a supernova. As it states here "Smiths cloud is probably one of the countless gas clouds that together have formed the Milky Way over billions of years" (Schilling 124).
- There are countless star clusters throughout the universe, "the arches cluster contains 150 massive bright stars in an area only two light years in diameter" (Schilling 125). Could you imagine that if Earth were closer to this star cluster it may never be dark

Intriguing Astronomical Scientific Information

outside, for the light from those other stars would illuminate the night sky.

- In the center of the Milky Way there is a monster black hole, "four million times more massive than the sun" (Schilling 127). This black holes mass is calculated by mapping the velocities at which the stars orbiting the black hole move. This process is made possible by using an infrared telescope.

- VY Canis Majoris is a star that extends 2000 times the size of the sun. That is the largest star we have found in the universe.

- Imagine a beam of light was bouncing back and forth between Los Angeles and New York City. The speed of a light is so great that it could do 38 bounces every second.

- Consider this; earth is orbiting the sun, while it is rotating on its own axis. The sun is orbiting around the Milky Way; the Milky Way galaxy is orbiting around our Local Group of galaxies. Gravity is the force that keeps this entire motion going.

- The Earth is orbiting the sun at about 66,000 miles per hour. The speed our sun is traveling through the Milky Way around Sagittarius A-Star, the black hole at its center is 483,000 miles per hour. The Milky Way itself is traveling through the cosmos at 1,300,000 miles per hour.

- Where the sun is located in the Milky Way the proximity of the distance between the stars is about 3 to 5 light years.
- The closest galaxy to the Milky Way is the Andromeda galaxy. This galaxy is around 2.5 million light years away. That means that it would take 25 Milky Way galaxies to connect the distance between these two galaxies.
- In the observable universe there appears to be something of more than 100 billion galaxies.

Intriguing Astronomical Scientific Information

GALAXIES

- In the constellation the big dipper (Ursa Major) there is a spiral galaxy. This galaxy is 30,000,000 light years away from our planet.

- Out of the 100 billion galaxies in the observable universe, astronomers have classified them into groups "these include majestic spiral and barred galaxies like the Milky Way, as well as colossal elliptical galaxies, misshapen irregular galaxies, dwarf galaxies, and starburst galaxies which contain enormously active star forming regions" (Schilling 145).

- Galaxies stick together in clusters throughout the universe. They are by no means a stone's throw away but, light years apart from one another. The large galaxies "are accompanied by a large number of small satellite galaxies, and they are almost always part of a group or cluster" (Schilling 145).

- "Practically all galaxies have a supermassive black hole at their core" (Schilling 145).

- Nearly all galaxies are categorized as either a spiral galaxy or an elliptical galaxy. Spiral galaxies are then divided into three types, each type is separated by how compact its stars are. "Spiral galaxies or (Type S) are comprised of three components, a flat rotating disk, which contains the gas and dust rich spiral arms, and an extended tail of dark matter and globular clusters" (Schilling 146).

- Elliptical galaxies or (Type E) are a large cluster of stars that are relatively near one another. They mainly just have stars and almost no interstellar gas. The movement of the stars in this type of galaxy is as follows, "the motion of the stars is less organized than in a spiral galaxy" (Schilling 146). Elliptical galaxies, have a lot of shapes and sizes, from spherical (E0) to very elongated (Eg). They vary in size from small dwarf galaxies to giant galaxies with many trillions of stars.

- "Lenticular galaxies (Type So) are an intermediate type between Spiral and Elliptical galaxies" (Schilling 146). This type of galaxy does have a spiral disk that consists of almost entirely of a flattened central bulge so that they are similar with that of Elliptical galaxies. Lenticular galaxies have used up or lost most of their interstellar matter, so they have stopped making new stars.

Intriguing Astronomical Scientific Information

- Irregular galaxies (Type Irr) are random clusters of only around somewhere of several 100,000,000 stars with no observable structure.

- The Pinwheel galaxy (M101) is in the constellation of Ursa Major, which is also called the big dipper. Ursa Minor is another name for the little dipper.

- The Whirlpool galaxy (M51) "Is an impressive two arm spiral at a distance of about 23,000,000 light years" (Schilling 150). I think it is really neat how when you look at this galaxy through an infrared telescope you can see how dust appears as darkness as it is distributed throughout its spiral arms.

- Galaxies can produce a large quantity of energy at several different wave lengths. "Such emission lines suggest the presence of large quantities of extremely hot gas" (Schilling 151). The super massive black holes in the center of these galaxies are believed to be surrounded by gas at extremely high temperatures and that is what powers all of the activity.

- It may be difficult for you to comprehend because of the reality of how tremendous it is but, "the observable universe has some 100 billion galaxies" (Schilling 154).

- "As so many new massive stars are born in M28, it should come as no surprise that many supernova explosions occur here" (Schilling 160). M28 is also called the cigar galaxy because that is what it is shaped like.
- Dwarf planets are really large asteroids. Brown dwarf stars are giant gas planets.
- In the constellation Virgo, there is an elliptical galaxy known as M87. Scientists have calculated the distance of this galaxy to be about 54,000,000 light years from planet Earth. Other than light waves and radio waves, M87 also emits a lot of gamma rays and x-rays. This galaxy is believed to be centered around a supermassive black hole. M87 contains a black hole that is 6.5 billion times as big as our sun.
- It is believed that the first galaxies would have been made nearly 100,000,000 years after the big bang. It is discussed here how these galaxies grew, "in the course of billions of years, the small galaxies merged to form larger ones. These often developed a majestic spiral structure" (Schilling 167). The big bang happened 13.77 billion years ago, just to put this time frame into perspective for you.

Intriguing Astronomical Scientific Information

- "High energy, short wave radiation from the universe (ultraviolet light, x-rays, and gamma radiation) can only be observed from space. Life on Earth is fortunately protected from these deadly rays by the atmosphere" (Schilling 168).

- In the late 1950s George Abell published the "northern survey". This was his catalog of the star clusters of galaxies that are visible in the northern hemisphere. This was "a catalog of 2712 clusters" (Schilling 171).

- Somebody later in the 1970s someone completed a "southern survey" which a listed another 1361 clusters that are visible south of the equator. These two books showed us that galaxies are not evenly spaced out throughout the universe. It showed us that "they are part of small groups that in turn form larger clusters, and these clusters are often times themselves grouped into elongated super clusters" (Schilling 171).

- The Virgo cluster is composed of at least 1500 individual galaxies. These galaxies are situated in the universe some 50 to 60 million light years from Earth. The Virgo cluster is the closest one to Earth. The Virgo cluster is about 15,000,000 light years across.

- The Milky Way is in the Local Group. The Local Group is part of the Virgo Cluster, and belongs to the elongated Virgo super cluster. The Virgo super cluster is composed of about 5000 separate galaxies. Today scientists believe that "the universe as a whole must contain around 100,000 super clusters" (Schilling 175). That is really unbelievably big.

- In 2016, the Spitzer space telescope revealed that the Andromeda galaxy contains about one trillion stars. This makes the Andromeda galaxy a little more than twice the size of the neighboring Milky Way. The Milky Way is believed to contain 200 to 400 billion stars. The Andromeda galaxy sits 2.5 million light years from earth. Andromeda is the largest galaxy in the Local Group. The Local Group also contains the Triangulum galaxy and around 44 other smaller galaxies.

- The Triangulum galaxy lies approximately three million light years from Earth, it's in the constellation Triangulum. The Triangulum galaxy is the most distant of galaxies that can be seen with the human eye. The Triangulum galaxy is also called the Pinwheel galaxy because that is what it resembles.

Intriguing Astronomical Scientific Information

- Our Hubble space telescope has found an estimated 100 billion galaxies in the visible universe. It is believed that there are at least another 100 billion galaxies in the universe. The location of these galaxies will have to remain a mystery, until our telescope technology in space improves enough to see them.

- The Milky Way galaxy is in the Local Group of galaxies. The Local Group of galaxies are just a part of the even larger Local Super Cluster (LSC) of galaxies. Scientists now know of many structures similar or even larger in size then the (LSC). These often times contain more than one group or galaxy cluster. These are called super clusters of galaxies.

- The (LSC) lies on the Virgo cluster, this is a grouping that is a cluster of galaxies containing several hundred bright galaxies that are a few mega parsecs (roughly 10 million light years) across.

DARK MATTER

- Scientists have discovered it is not what shines in the light but, what hides in the dark that holds the secrets of our sky.
- There is a mysterious dark matter that binds stars and galaxies together.
- Scientists have realized that dark matter exposes itself by bending light that passes through it, this act is called gravitational lensing.
- Dark matter particles are traveling at the speed of light, and they don't interact with anything well.
- Dark matter is not affected with real matter. Scientists have come to find out that billions of particles of dark matter are passing through the Earth every second of every day.
- At the moment of the big bang dark matter was created, and it played a critical role in helping ordinary matter clump together to form stars and planets.
- It is estimated that dark matter makes up 23% of the universe but, ordinary matter makes up only 4%. Scientists were amazed to find that the other 73% of the universe was dark energy.

Intriguing Astronomical Scientific Information

- Science has always assumed that even though the universe continues to grow in size, it would eventually slow in its expansion. Or it was thought that the universe would stop growing and collapse on itself. Scientists were shocked to find the expansion wasn't slowing but speeding up. Eventually the universe will become so cold that every living thing will freeze to death. Of course, this will be subsequent to the expiration of all of the clouds of gas that are active with stellar formation.
- The greater the distance a galaxy is from Earth, the greater the speed it is moving away from us.
- The expansion of the universe only applies to objects that are not bound together by gravity. For example the Earth is not getting any closer or farther than normal from the sun as time goes on. So as long as the gravity of our galaxy keeps pulling, it will hold us in place.
- Strange particles exist in the universe, they are called W.I.M.P.S, axions and monchoes. These particles are assumed to help dark matter bind things together. Combined dark matter and dark energy make up 90% of the universe.

- Science has not directly proven that dark matter particles really exist. This is due to the fact that dark matter doesn't emit light and it doesn't absorb light either. Dark matter doesn't interact with the light at all. Almost every textbook on the planet tells us that our world and universe is made out of atoms and some subatomic particles. On the contrary though if dark matter really exists, then that would introduce a new way to consider how things are. Dark matter is believed by scientists to surround galaxies and to hold them together. They say that it is… like water in a fish tank it holds things at certain levels and gives them a substance to order their centers of gravity around. It keeps the rotation of all of the matter in a galaxy at a consistent speed.
- Most of the galaxies mass is from its invisible dark matter that surrounds all of the regular matter like water in a fish tank.
- Ma.C.H.O. (massive compact halo object) these are high mass objects that do not emit much light and are present throughout the galaxy.
- There is believed to be at least 10 times as much dark matter as regular matter that exists in the universe.

Intriguing Astronomical Scientific Information

- Many scientists believe that dark matter is a new exotic particle and billions are passing through us at every second.
- W.I.M.P. (weakly interactive massive particles) these are nearly the perfect candidate for what dark matter could be.
- Here is why scientists are searching for dark matter, they believe it will show them what was going on one $10,000^{th}$ of a second after the big bang.
- Before the big bang there is no center you could point to. There was no direction in the sky because there was no sky.
- Dark matter is thought to have formed a web like structure that ordinary matter has clung to, to form galaxies.
- Scientists have learned that galaxies just don't randomly appear in space, they needed dark matter to give them a structure to form around.
- Dark matter is not believed to be something you could hold in your hand, for it is thought that it would sink right through it and go into the ground.
- When light goes through dark matter it bends, just the way light bends when it goes through glass. This feature of dark matter has clued scientists into its existence.

- It is stated that dark matter makes up 23% of the universe, where ordinary matter takes up only 4% of it. Dark energy is believed to take up all of the remaining 73% of the universe.
- If you look at progressively more distant galaxies, you are seeing them as they were at progressively greater times in the past.
- A supernova is the colossal explosion of a star at the end of its life. This occurs when a dying star known as a white dwarf goes through a nuclear runaway. It literally blows itself to smithereens.
- Dark energy has a repulsive effect that dominates over gravity and this means galaxies are accelerating away from each other faster and faster as the universe grows.
- Dark energy is the energy of the vacuum of outer space. Even nothingness has energy and it is pushing the galaxies apart creating a runaway universe. Dark energy is creating and therefore expanding outer space.
- It looks as if dark energy, dark matter and the laws of physics are a death warrant for all intelligent life in the universe.

Intriguing Astronomical Scientific Information

THE LOCAL GROUP

- The Local Group contains three larger galaxies, each containing billions of stars. The Milky Way galaxy is "together with the Andromeda galaxy in the slightly more distant Triangulum galaxy and a number of smaller galaxies, make up the Local Group" (Schilling 135). With the three heavy weight galaxies plus each of the smaller dwarf galaxies, the Local Group has more than 50 members.

- There are an extraordinary amount of galaxies in the visible universe. Most of which are small dwarf galaxies. As gravity continues to pull on matter, galaxies floating in outer space will continue to grow in size by swallowing up smaller galaxies. The entire Local Group is being pulled towards the Great Attractor, which is the largest known galaxy in the universe.

- The large Magellanic cloud contains a few billion stars. It has been believed to have once been a barred spiral galaxy, just like our own galaxy. Although these galaxies are pulling on the large Magellanic Cloud "have been strongly distorted by the tidal forces of the Milky Way. The cloud is about 167,000 light years away" (Schilling 137). This region of space contains an extremely large amount of stellar gas and dust that "the level of star forming activity is therefore much higher than in the Milky Way" (Schilling 137).

- A Cepheid is a star "that gets fainter and then brighter again over a period of a few days or weeks" (Schilling 138).
- Astronomers use the Leavitt law or what is known as the period luminosity relation, this can help them to determine the distance to other stars and galaxies.
- The Andromeda galaxy is one of the most distant objects that can be seen in the universe without optical instruments. It is seated in the universe "around 2.5 million light years away, it is just visible with the naked eye, as a faint smudge of light to the northwest of the star Nu Andromedae" (Schilling 141).
- Hundreds of years ago galaxies were believed to be separate island universes.
- The Spitzer space telescope has revealed that the Andromeda galaxy contains over one trillion stars, more than twice as many as the Milky Way.

Intriguing Astronomical Scientific Information

- The Triangulum galaxy is between the constellations Andromeda and Aries. Under extremely clear and dark conditions it can be seen with the naked eye. Doing this will require perfect eyesight. The Triangulum galaxy is the smallest of the three large billion star galaxies in the Local Group. It is only half the size of the Milky Way, it contains 40 billion stars. The Triangulum galaxy is a spiral galaxy but, does not appear to have a black hole in its center.

LIGHT-YEARS (LY)

- Sure everyone knows at light speed is fast but, let me put a number to those words. Light travels 186,000 miles in one second. When you consider how many seconds there are in a year, then multiply those two numbers, this is the distance of one light year. Here is the distance of one light year, it is almost six trillion miles long.

- Time is believed to slow if someone were to travel at the speed of light. The speed of light is equal to this equation (3.0×10^8 m/s). Albert Einstein was the first physicist to come up with the "Theory of Relativity". He proposed this equation in 1905 ($E = mc^2$). In this equation E stands for energy, M stands for mass and c^2 stands for the speed of light squared. In this theory he says that the speed of light always remains constant, regardless of the speed of the observer.

- It takes one and a half seconds for us to see light reflected off the moon. It takes light from the sun 8 or 9 minutes to reach the Earth.

Intriguing Astronomical Scientific Information

- The distance to the moon is roughly 250,000 miles away from the Earth. The sun is 93,000,000 miles from Earth on average.
- The speed of light is also called an angustrum.
- The Milky Way rotates once every 250,000 years.
- The closest star to the sun is 7000 times further than the planet Pluto. Let me change that 7000 times further than the rock Pluto. The reason I made this change is because some scientists have questioned Pluto's status as a planet. This is due to it being just a piece of rock that was captured by our sun and cast into orbit. This rock is believed to have once been a part of the Keiper belt.

SUPER CLUSTERS

- Super Clusters are large groups of smaller galaxy clusters or galaxy groups and are the largest known structures in the universe. The Milky Way is in the Local Group of galaxies, this group of galaxies lies in the Laniakea Super Cluster.

- Astronomers believe that there are some 10,000,000 Super Clusters in the observable universe.

- Laniakea Super Cluster Spans over 500,000,000 light years, while the local group Spans over just 10,000,000 light years. Galaxies are grouped together to form clusters instead of being disbursed randomly in outer space. Clusters of galaxies have been grouped together to form Super Clusters.

- Super Clusters are made of dozens of separate clusters throughout an area of space that spans close to 150,000,000 light years across.

- Super Clusters are not bound together by gravity; galaxies are free to drift around freely.

- The biggest galaxy in the universe is called the Great Attractor. This galaxy contains so much gravity it is pulling all of the

Intriguing Astronomical Scientific Information

galaxies within the Local Group toward it, at an astounding rate of speed. This includes the Milky Way at a rate of several hundred kilometers per second.

- The Local Super Cluster has a diameter of 100,000,000 light years. It is also called the Virgo Super Cluster.
- The largest Super Cluster that exists in the universe is the Perseus Pegasus Filament. The Perseus Pegasus Super Cluster extends one billion light years and is the largest known structure in the visible universe.
- Here is a list of galaxy Super Clusters we currently know exist in the visible universe, Laniakea Super Cluster, Virgo Super Cluster, Hydra-centaurus Super Cluster, Perseus Pisces A Super Cluster, Pavo-Indus Super Cluster, Coma up Super Cluster, Sculptor Super Cluster, Hercules Super Cluster, Leo Super Cluster, Ophiuchus Super Cluster and the Shape Super Cluster.
- Then you have the distant Super Clusters which are: Pisces-Cetus Super Cluster, Bootes Super Cluster A, Horobgium Super Cluster, Corona Borealis Super Cluster, Columba Super Cluster, Aquarius A Super Cluster, Aquarius B Super Cluster, Aquarius Capricornus Super Cluster, Aquarius cetus Super

Cluster, Bootes Super Cluster B, Caelam Super Cluster, Draco Super Cluster, Draco visa major Super Cluster, Fornax-Eridanus Super Cluster, Grus Super Cluster, Leo A Super Cluster, Leo sextans Super Cluster, Leo Virgo Super Cluster, Microscopium Super Cluster, Pegasus Pisces Super Cluster, Perseus Pisces Super Cluster, pisces Aries Super Cluster, Ursa majoris Super Cluster, and the Virgo coma Super Cluster.

- There is yet one last grouping of far distant Super Clusters. These are the lynx Super Cluster than it goes to numerical codes and those are like, SCL@1338 at z=1.1, SCL@1604+43 at z =0.9, SCL @0018+16 at z =0.54 in SA26 and MS0302+17.
- Now that you have heard all that we know to be in existence at the writing of this book, hopefully you will have a newfound respect for gravity because of the motion it has caused and continues to have on matter in the universe.
- I am confident that mankind will continue to expand our knowledge about outer-space because that's the nature of our species, to improve upon what we have.

Intriguing Astronomical Scientific Information

CONSTELLATIONS CASSIOPEIA

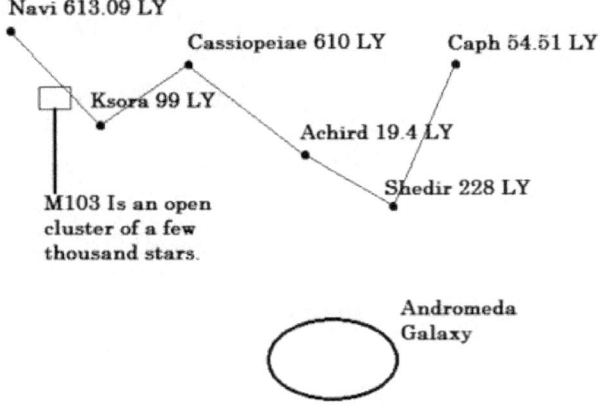

- Cassiopeia- is a constellation in the northern sky. That is named after the vain queen Cassiopeia from Greek mythology, who boasted about her unrivalled beauty.

CONSTELLATIONS
THE BIG AND LITTLE DIPPER

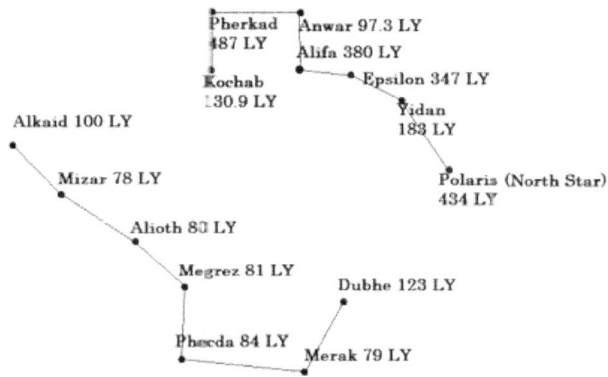

- The Big Dipper has an astrological name, which is Ursa Major. It also helps to make the constellation called The Great Bear. The two stars that make the front of the big dipper (Dubhe and Merak), will get you pointed in the right direction when looking for the North Star Polaris.

- The Little Dipper is also part of the constellation Ursa Minor (The Little Bear).

Intriguing Astronomical Scientific Information

CONSTELLATIONS
ORION

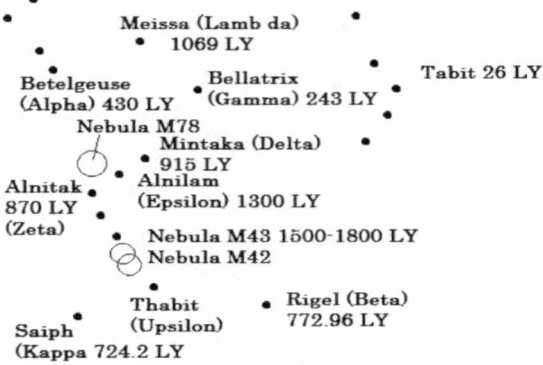

- One story is that Orion inherited his great hunting talents from his mother. She was Queen of the Amazons and was a well-known great huntress. Orion angered the god Apollo by bragging about being the best hunter in the land, who could kill any animal in the world. This made Apollo send a scorpion to sting Orion on the foot killing him.

- Nebulae appears as an indistinct bright patch or as a dark silhouette against other luminous matter. They are massive clouds of dust, hydrogen, helium gas, and plasma , they are also

"Stellar Nurseries". Under extremely dark conditions the Nebula M78 is visible to the naked eye.

Intriguing Astronomical Scientific Information

CONSTELLATIONS
THE SEVEN SISTERS
(Pleiades Cluster)

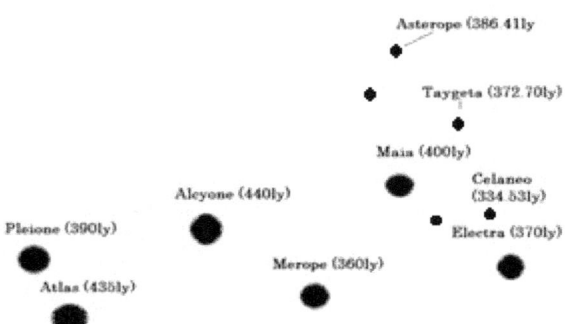

- The story of this constellation is there were seven sisters Maia, Alcyone, Asterope, Celaeno, Taygeta, Electra and Merope. Their parents were Atlas and Pleione. Atlas was commanded by the god Zeus to hold up the earth, while Pleione was the mythical protectress of sailors.
- Here is a tip for locating the seven sisters in the night sky. First find the constellation Orion. Then use the three stars that make his belt. Make an imaginary line going through them. Extend this line upward through a cluster of stars, continue this line

further upward in the same direction. The stars in this constellation are not bright but, always remind me of a kite.

Intriguing Astronomical Scientific Information

TIME TRAVEL

- In Einstein's theory of relativity, time plays a central role. In this theory time travel remains as an open question. Traveling into the past seems improvable because us humans have only experienced time moving in one direction "forward". Physicists call this the arrow of time.
- The universe has a law that goes "all things go from orderly states to disorderly states".
- One theory of time travel is that there are multiple parallel universes. Time travel in that sense would be like saying you entered another universe at another place in time. So you wouldn't affect the universe from which you came.
- Us humans think that nature always manages to choose the solution that provides self-consistency.
- Another theory of time travel would require someone to travel in a space ship and park it very close to a black hole. The immense gravity of the black hole would slow time. This person in the space ship would return to Earth to find that

many years have passed on Earth, where they would think that their trip only has lasted a couple weeks.

- Alpha Centauri is the nearest star to the sun. It is 4.3 (ly) away which calculates to 25 trillion miles. That is 100 million times further than the moon.

- A worm hole is a theory, in which it acts as a gateway from one place in the universe to another. In this theory you enter the worm hole and come out at a completely different location, possibly (ly's) apart from one another.

- Getting close to a light speed appears to be the most promising way to travel in time, by slowing down our clocks. If this is possible we have yet to find a way to do it .

- When astronomers recently discovered that the expansion of the universe is not slowing but, accelerating they reasoned that it was driven by the mysterious dark energy. We don't really know what this dark energy is but, we have assumed that it is a property of outer space that cannot be changed.

- Scientists and astronomers hope to one day discover an Earth like planet in rotation around the star Alpha Centouri B. This star is believed to have a habitable zone similar to our sun.

Intriguing Astronomical Scientific Information

BLACK HOLES

- Black holes have so much gravity that this fact alone makes them the most destructive objects in the universe.

- Astronomer Michelle Thaller says this about black holes, "they are sort of cosmic organizers, they bring things together."

- As something falls into a black hole it increasingly gains speed faster and faster until it reaches the speed of light. No one knows what happens to this object once this happens. We know that if we can understand this, it will allow us to understand the nature of relativity better.

- Sagittarius A-Star is the black hole that our galaxy the Milky Way is centered around. It is a super massive black hole four million times the mass of our sun. This monster sits 26,000 light years from Earth. As earth revolves around our sun, it takes around 250,000 years for our sun to make one lap around our black hole.

- G2 is an interstellar cloud of dust and gas that scientists believe holds a star. G2 is very close to Sagittarius A-Star. Parts of this cloud will be consumed by the black hole while other parts will be catapulted into space.

- Inside the galaxy NGC 4889, 300,000,000 light years away, is the largest black hole ever found.

- Friction of particles near a black hole causes them to generate so much heat that they begin to glow.

Intriguing Astronomical Scientific Information

THE SUN

- Our Sun is 4.6 billion years old it has been burning in the same state for the last 4 billion years and is expected to remain as is for the next 5 billion years. The Sun is approximately 93,000,000 miles from the earth. The sun converts Hydrogen into Helium. Our Sun is pretty average when compared to other stars. The Sun is 109 times as wide as the Earth.
- There are 3 types of stars; they are Super Giants, Main Sequence and Dwarfs.
- The Sun generates its heat by creating nuclear fusion in its core converting H to HE.
- The sun is a Main Sequence star. It is believed by most astronomers to have a Main Sequence lifetime of approximatley 10 billion years.
- The atmosphere of the Sun is composed of several layers, mainly the photosphere, the chromosphere and the corona. There are solar flares that burst through these atmospheric layers, only to be pulled inward by the Suns gravity. From the center of the Sun out, its composition is divided into a core,

followed by the radiative zone, and then the convective zone. Above the convetive zone are the atmospheric layers of the Sun.

Intriguing Astronomical Scientific Information

TRAPPIST-1

- On 2/22/2017 the Spitzer Space Telescope, which is an infrared telescope, revealed a batch of seven Earth-size planets in orbit around a single star.

- Trappist-1 is an ultra-cool dwarf star, located 39(ly) away from planet Earth. From our vantage point here on Earth the star sits in the constellation Aquarius. This star is $1/8^{th}$ the mass of our sun and is only about half its temperature.

- Each of these planets orbit Trappist-1 closer than Mercury is to the sun.

- The density of all of these planets signifies that they could all be rocky worlds and given their cool temperatures, this indicates that with the right atmospheric conditions, all could host liquid water on their surfaces. These are two of the main factors for life as we know it to exist. Three of these seven planets are located in the stars habitable zone, as they are the ones most likely to have liquid water on their surfaces.

- This other solar system is 235 trillion miles or 40 (ly) away from planet Earth.

- These seven planets are so close to one another that in some cases, a neighboring planet might appear twice as large as the full moon is seen here on Earth. You would be able to see structures on the other worlds. If they have sky scrapers like us here on Earth.
- Trappist-1 is 2000 times dimmer than the Sun and just slightly larger than the planet Jupiter.
- The habitable zone of Trappist-1 is very close to the star. That is the region around a star where the planet could have a surface temperature right for liquid water.
- "Ultracool dwarf stars produce significantly less radiation then Sun-like stars and most of Trappist-1's light is radiated in the infrared wave lengths rather than visible wavelengths".
- Since the light radiation from this star is released in infrared, if there are alien creatures on these planets would there vision be based in infrared? This will remain a mystery until contact is made.
- Prior to the discovery of this solar system our scientists just had one solar system to study, Mercury, Venus, Earth, and Mars. "Now we have seven more that we can study in detail.

Intriguing Astronomical Scientific Information

- Red dwarf stars burn for such a long time it has been thought that some have eternal life. Trappist-1 is believed to be a fire that will burn for trillions of years. It will be burning long after the Sun goes dark.

- In order to tell the approximate lifetime of a star scientists take the luminosity of a star then divide it by C^2. This gives you the mass rate of change. Then divide the mass of the star.

CONCLUSION

- As I mentioned earlier gravity is the culprit that has caused motion in the universe. All of the atoms that make up all of the gases, liquid and particles of matter in the universe are floating in outer space. Here's a quick look at every piece of substance we know to be in existence. Our home planet Earth, than our solar system, (consisted of the Sun and all eight planets). Then it is our solar interstellar neighborhood (or the stars that are closest to us). Next comes the Milky Way galaxy, followed by our Local Galactic Group, after this comes the Virgo Super Cluster. Finally comes the 25 Local Super Clusters and all of this is just part of the observable universe.

- Here is one theory of what will become of the universe and of the atoms that make the gases, liquid and the particles of matter in it. In this scenario, everything that has happened since the big bang is all part of an extraordinarily long reoccurrence. That is everything that is in the universe has already happened before and will continue to happen time after time again. That would mean that there has been a countless number of big

Intriguing Astronomical Scientific Information

bangs. The gravity that has caused motion by pulling on things, just continues to pull on things and keeps the cycle repeating itself time and time again.

- The first scenario is actually a lot nicer then the second scenario. Unfortunately though the second scenario already appears to be happening.

- In the second scenario everything that is in the universe continues to drift from its current position in the universe. That is the universe continues to expand in every direction above, below, to the left, to the right, forward and behind with every passing second. This movement speeds up as the distance increases. That is, the further a galaxy is from us the greater the speed that it is receding away from us.

- So as I said before in this scenario, that as all of the stars that are in the universe burnout and die all of the galaxies will continue to get further and further apart. Everything will be trapped in an eternal deep freeze, it will be pitch black everywhere and there'll be no light or heat; as all clouds of interstellar gas have become exhausted and have stopped making new stars.

- When you look at the night sky from anywhere on Earth, "the stars and galaxies that we see in our universe constitute for only 1% of the total matter energy content" (Schilling 197). The entire universe is inconceivable for the human mind to grasp. Its size is so extensively spacious, that the end it seems will never be found.

- When you consider just how immense our universe is, take this into consideration. "Scientific journals are full of speculation about an enormous range of possible parallel universes" (Schilling 201). If there are other universes, could there possibly be a gateway connecting them? As far as those thoughts may be, we humans have enough trouble getting along with each other! It does not appear that we could coincide well with an alien species and exist with each other in tranquility. When we can live together in peace and share commodities, food, land and resources with each other, then we may be ready to coincide with an alien species. Considering they are peaceful and have good intentions of sharing the universe with us.

- Here is a quote regarding the likelihood of another Earthlike planet existing in orbit around a star similar to our sun.

Intriguing Astronomical Scientific Information

"Astronomers estimate that one in five sun like stars is accompanied by a small rocky planet in the habitable zone" (Schilling 81). Given this information, it only feeds our desire to keep hunting to find the real twin sisters of the Earth. We have been looking but the first is yet to be discovered.

- Scientists presume that after its ten billion year life as a main-sequence star our Sun "will swell up to become a red giant and blow a colorful, expanding planetary nebula into space" (Schilling 83). Once it reaches the extent of its red giant phase and releases its "planetary nebula" into space, "it will contract until it is a small, hot white dwarf that will get fainter and cooler, like a gradually dying cinder" (Schilling 83). A stars "planetary nebula" it releases into space, is captured by the gravity that each atom in outer space has and combines to form new stars and planets. So if the universe is similar to a recycling plant then in that sense, it will keep renewing itself time and time again. This process is so long that one would never know if it were to be true or not but, here's to hoping it is anyhow.

Works Cited

Murphy, Edward M. Ph. D. Our nights sky. Course Guidebook. Chantilly, Virginia: The teaching company, 2010.

Schilling, Govert. Deep Space. Beyond the solar system to the edge of the universe and the beginning of time. New York, NY 10011: Govert Schilling and black dog & Leventhal Publishers inc., 2014

Sparrow, Giles. Hubble. Legacy edition. New York: Metro Books., 2015.

Television Series: Strip the cosmos, cable, SCI HD., Ch. 1132

Television Series: The universe, Cable, HdH2. Ch 1136.

Trefil, James. Space Atlas. Mapping the universe and beyond. Washington, D.C. 20036-4688 USA: National Geographic Partners, LLC. 2012

www.ingramcontent.com/pod-product-compliance
Lightning Source LLC
Chambersburg PA
CBHW070145230526
45471CB00002B/531